David Heaf

Natural Beekeeping with the Warré Hive
A Manual

Northern Bee Books

Published by:

Northern Bee Books
Scout Bottom Farm
Mytholmroyd
West Yorkshire HX7 5JS
Tel: 01422 882751
Fax: 01422 886157
www.GroovyCart.co.uk/beebooks

ISBN: 978-1-908904-38-6

Printed by Lightning Source

Cover Design: D&P Design and Print

Front cover: Warré hives in a Perthshire garden. Photo: Andy Collins
Back cover: Warré hive comb. Karman Csaba, Romania

Contents

Preface

This book is for people who need no convincing that *natural* beekeeping is for them, and would like to try it out with the Warré hive, either as a first hive or after having used another. It contains all you need for getting started in beekeeping with the Warré hive and managing it in later seasons once the bees are established. It contains very little on the philosophy, ethics and science behind natural beekeeping. I cover those aspects in detail in my book *The Bee-friendly Beekeeper*[1] . Furthermore, I assume that the reader will familiarise themselves with basic bee biology and behaviour by reading the excellent and colourfully illustrated books on the subject, a suggested list for which I provide in §1.6. The story of how and why the Warré hive, called by the inventor, 'The People's Hive', came into existence is told in detail in Abbé Émile Warré's *Beekeeping for All*,[2] the last original French edition of which was published in 1948. Whilst the present book is essentially a complete guide including much practical detail that is in neither of the aforementioned books, I thoroughly recommend Warré beekeepers to read Warré's original book, which was ahead of its time in approaches to sustainable, natural beekeeping.

Since the translation of Warré's book into English in 2007 the People's Hive has spread to nearly every continent, and the hive is now used in rural and urban settings from the tropics to the taïga. I therefore try to accommodate in this book the increased diversity of surroundings in which the hive is used, drawing on reports from Warré beekeepers worldwide. There are a few procedures described which I have not needed to use myself, but which I feel should be included for the sake of completeness.

It is worth keeping in mind that honey bees are amazingly adaptable, settling in cavities of all sorts of shapes and sizes, and sometimes in the open. I myself have relocated nuisance colonies from cavities ranging from water company valve chambers in the ground, to chimney pots on three-storey buildings. This means that bees are not very choosy about the type of hive available as long as it has sufficient volume to form a viable brood nest. Among the more natural types of hive, we can list: skeps, logs, horizontal (Kenyan/Tanzanian) top-bar hives, the sun hive (a two-piece skep with rounded frames), and vertical top-bar hives of which the Warré hive is one. Its relative naturalness derives mainly from the fact that the combs are foundationless and, unlike those in frames, are fixed to the roof and walls of the hive, as happens in honey bee nests in natural cavities. Also, the nest is allowed to grow downwards, as happens in a hollow tree trunk. Much of the rest of the naturalness of the Warré hive is derived from the way it is managed. Therefore, it must be conceded that even a Langstrothian type of hive could be managed relatively naturally, for example by using foundationless, i.e. near-natural comb[3] . However, were a natural beekeeper to require the use of frames, for example to comply with state regulations, Warré developed a frame version of his hive which I briefly touch on in

---

[1] Northern Bee Books, 2010
[2] Northern Bee Books, 2007/2013; transl. by David & Pat Heaf of *L'Apiculture pour Tous*, Émile Warré, 1948.
[3] Fully natural comb allows the bees to determine cell size *and* comb spacing.

this book. Other more natural frame hives include the golden hive (*Einraumbeute*, one-box hive), a trough hive with very deep frames that avoid horizontal interruptions in the brood nest, or the somewhat more challenging sun hive (Weißenseifen Hanging-basket hive) already mentioned.

Readers may have noticed that the term 'natural beekeeping' is an oxymoron: once you put bees in a container you have taken the first step away from naturalness. But using the term 'relatively natural beekeeping' would be cumbersome. Maybe 'apicentric beekeeping' or 'bee-centric beekeeping' are more appropriate terms, but the term 'natural beekeeping' seems to have been accepted by the *Zeitgeist*. Indeed, in the UK there is a 'Natural Beekeeping Trust'[4] and the term appears in the title of beekeeping books and web pages. Therefore what is meant by 'natural beekeeping' is now adequately established, so we will stick with the term, despite its contradictions. If Langstrothian beekeeping is considered modern, then natural beekeeping is post-modern.

Just as beekeepers in general have fundamental attitudes which cover a spectrum from the more anthropocentric to the more apicentric, so too among natural beekeepers there is a range of attitudes. Although I confine myself to one type of hive in this book, I discuss some interventions which might make more radical natural beekeepers raise an eyebrow. Yet Warré too included some more anthropocentric manipulations – for example, swarm control by splitting the hive. Even so, he did not explicitly claim to be presenting natural beekeeping. At the other extreme, especially if you live in a rural area, you may leave your hive entirely alone, treating it as if it is a hollow tree, and visiting it only to marvel at the life of the bee. Subject only to beekeeping legislation, the choice is yours to do what works for you.

Natural beekeepers may like to claim that their bees are healthier than those kept by more artificial methods. There is much in the apiological literature that could be adduced to support this view, and I have covered some of it in *The Bee-friendly Beekeeper*. However, until long-term controlled experiments are carried out in different climatic regions involving a comparison of the different hives and methods of management, it is not possible fully to substantiate that broad a claim. Here is an opportunity for apicultural researchers worldwide!

David Heaf
March 2013

---

[4] www.naturalbeekeepingtrust.org

1   Introduction

1.1 Natural beekeeping

In the following brief listings of possible criteria for natural beekeeping some may have to remain as unachievable, at least for the time being.

1.1.1 Hive and site

- wood and/or natural fibre with no metal or plastic
- shape suited to the shape of a swarm or cluster
- scope for vertical or horizontal colony growth
- volume and elevation above ground as found when bees offered the choice
- entrance size, orientation and position as found when bees offered the choice
- wall thickness as in hollow trees
- uninterrupted brood nest
- queen is in principle free to roam throughout; no excluder
- shaded from hot sun as in a tree trunk
- not in a wet dark hollow
- forage accessible and sufficient for the colony density desired, including wintering
- no preservatives on the hive or stand; innocuous natural paints only

1.1.2 Management

- bee-centred instead of maximising production of hive products for profit
- natural comb; no foundation
- comb fixed to top and walls of hive cavity; no Langstrothian 'bee spaces'
- absolute minimum interference with the hive
- very rarely opening the top, which lets the heat out
- supering only where there is a risk that honey binding compromises wintering
- no or minimal use of smoke
- colony reproduction by natural swarming
- no queen mutilation (marking, wing clipping)
- no artificial queen breeding, transplanting larvae, insemination
- locally adapted bees if available
- no chemicals in the hive that the bees do not put there themselves, e.g. no varroa treatments
- no drone culling
- hive locality free of pesticide use
- feeding only where essential and then only honey[1] and/or real pollen
- consider impact on other pollinators

What a lot to live up to! Clearly some compromises have to be made at the outset to make keeping bees possible. For example, thick walls, or putting hives at a height above ground more commonly chosen by bees, namely five metres would make handling boxes difficult and unsafe. And beekeepers in their first season may not

---

[1] See caveat § 11.2.1

have their own honey to feed in an emergency or, having only one hive, may not wish to leave their bees entirely to the ravages of varroa and the viruses it carries. So the above lists are not meant as some absolutist dogma, but rather as suggestions. Natural beekeepers come in various shades of ecocentricity and apicentricity. I suggest that we keep bees in a hive that suits us and in the way that satisfies us. So feel free to pick and choose!

1.2 Why Warré?

After four years of using a Langstrothian-type hive called the 'National' or 'British Standard', a few of which I still use, I came across Johann Thür's book praising an 18th century hive by Johann Ludwig Christ which comprised stacked boxes with top-bars in each box. To me its attractive feature was that the combs were fixed to the walls and the top of the hive, creating cul-de-sacs which retained the nest scent and heat (*Nestduftwärmebindung*). Furthermore, the hive was extended at the bottom as the colony grew, and boxes of comb were harvested from the top as they filled with honey. The comb was not reused. I then came upon the identical principle in the 'People's Hive' in a French book by Abbé Émile Warré which set out its construction and management in detail. I have since made and populated sixteen of these hives while reducing my Nationals to less than a handful.

What I like about the Warré hive is as follows:

* retention of nest heat and scent; better control of temperature and humidity
* thick, vapour permeable 'quilt', provides top insulation and possibly buffers humidity
* top-bar cloth replaces the crown board (inner cover); no cracking propolis; gentle entry
* smaller, more manageable and easier to construct boxes
* all boxes the same size
* comb built freestyle regarding cell-size and proportions of worker/drone cells
* brood nest constantly moving down onto fresh comb, boxes of old comb removed complete with honey from the top
* comb renewal by the bees and thus disease minimisation is automatically an intrinsic part of the annual cycle
* no costs of frames or foundation, or of the work involved in preparing them
* no foundation means that no contaminated wax is introduced
* no queen excluder
* no framed comb to extract, just cut and drain, so no expensive extractor, just kitchen utensils
* normally the hive need be opened in the strict sense only once a year at harvest; at the spring visit boxes are added underneath without letting out hive heat or disturbing the bees
* sturdy projecting box handles lend themselves to use of a do-it-yourself hive lift

But let us put all our cards on the table and look at some of the potential or alleged disadvantages:

* cavity not cylindrical like a hollow tree or a skep; Warré acknowledged a cylind-

drical shape as desirable but not easily practicable
- brood nest interrupted by top-bars
- more care needed when inspecting combs compared with Langstrothian hives
- honey harvested from brood comb; higher content of brood comb substances including pollen
- little or no reuse of comb so more nectar needed to replace comb harvested, therefore lower honey yields
- splits, if any, are more difficult to perform without frames
- more risk of honey contamination by varroa treatments, if they are used
- mentoring is harder to come by in the immediate locality of a beginner
- more challenging to assess colony condition by observing from without

Well, we can't have everything! We shall discuss some of the potential disadvantages later in the course of the book.

1.3 Mentoring and other sources of practical advice

In the last few years, courses on natural beekeeping and the Warré hive have been held in the USA, UK and Australia. With natural beekeeping currently still growing it should be possible to locate these courses through beekeeping publications and forums. The Natural Beekeeping Trust (UK) is devoted to natural beekeeping in general and runs courses.

Finding a nearby mentor for the Warré hive may still prove difficult. In any case, it is prudent to join one's local beekeeping association. Their meetings, especially those in apiaries in spring and summer, are highly informative, even if you don't always approve of their methods. Mentoring by an experienced beekeeper who is committed to Langstrothian hives, and nevertheless open-minded, can be an enjoyable experience. The understanding and sensitivity for bees brought by years of working framed comb can largely be transferred to dealing with bees in Warré hives. Less commonly, you might be met with a degree of hostility in your local association when your intentions and beliefs are disclosed. In which case it is better just to look, listen and learn.

Many beginners have obtained their mentoring from like-minded beekeepers a considerable distance away, either by phone or through the internet. This often happens via the many online forums, the main English one devoted to Warré beekeeping being the Yahoo e-group 'warrebeekeeping' which responds to queries, often within hours.[2] Web links to other forums are given in Appendix 1.

In the UK, the National Bee Unit, and in the USA, various state extension services, provide advice to beekeepers, particularly in relation to disease and pest management. Such services often have a network of local officers. The wealth of knowledge held by the various governmental services is not to be dismissed, even if it sometimes seems based on a world outlook into which natural beekeeping does not easily fit. In contrast, my local inspector used to run skep hives which he made himself. Hopefully, regarding the official information, you will gradually begin to appreciate what to take and what to leave. As inspectors vary in how they interpret their duties you may wish to familiarise yourself in advance with your own rights

---

[2] http://uk.groups.yahoo.com/group/warrebeekeeping/

and the code of practice or guidelines that should be implemented by the civil servant inspecting your hives. In the UK, such codes/guidelines are detailed on the National Bee Unit website.[3] In § 7.2.2, I suggest a way of working with your inspector that may be open to you.

1.4 The law and bees

Some countries or states require beekeepers to be registered. In many places there are legally notifiable bee pests and diseases, such as foulbrood. The regulations are obtainable from country/state agriculture departments. Your local beekeeping association will put you in touch.

Some local government administrations have bylaws or ordinances related to keeping bees, sometimes prohibiting apiaries in certain areas or prescribing separation distances from public thoroughfares. Again, your local association can advise.

1.5 Beekeeper health, safety and insurance

How you get on with your bees depends a lot on your behaviour towards them and that in turn has a lot to do with your inner attitude. A planned, purposeful, unhasty, calm, aware, and almost contemplative approach is more likely to be rewarded with co-operation than the opposite extreme, which I will leave you to imagine. But even with the best will in the world, you may have to sustain occasional stings. If they are frequent, consider what may be amiss: wrong time of day; colony already much disturbed by you or an animal (woodpecker); queenlessness; the perfume you are wearing; opportunist bees from other colonies coming to steal honey while the hive is open, and so on.

The main hazards in beekeeping are stings to oneself and others, and back injury. Beekeepers get used to occasional stings and are no more bothered by them than those from a nettle. However, rarely, a sting can be fatal, especially through resulting anaphylactic shock. Until you know your reactivity to stinging, it is prudent not to work alone. Sometimes, even a beekeeper with long exposure to stings may later develop a severe reaction. When stung, it helps to minimise the resulting inflammation if the sting is immediately scraped from the skin by a sideways movement of the edge of a hive tool, fingernail or bank card. The British Beekeepers Association has produced an excellent leaflet (L002) on stings which covers first aid in anaphylaxis.[4] A sting to the eye could cause blindness. For that reason, I always wear a veil when working with bees.

When bringing others to your bees, it makes sense to provide them with at least veil protection and make yourself familiar with the first aid for stings. You will no doubt avoid handling your bees if there is an obvious risk to family, neighbours or passers-by. Many beekeepers in the UK, possibly a majority, have public liability insurance, which is generally obtainable through their local beekeeping association, at least in the UK. It also covers liability in the unlikely event of a product, e.g. honey, proving harmful. I personally know of no beekeepers taking out insurance for loss or damage of hives and colonies, but am aware that it is available in France at a low cost per hive.

---

[3] https://secure.fera.defra.gov.uk/beebase/
[4] http://www.bbka.org.uk/files/library/bee_stings-l002_1342858887.pdf

Another kind of insurance available in the UK through beekeeping associations is bee disease insurance. It indemnifies against loss of frames and combs if they have to be burnt by order because of foulbrood infection. Other parts of the hive are generally sterilised by scorching with a blowtorch, or, in some countries, by irradiation. Warré beekeepers would suffer a much smaller loss in such circumstances as they do not invest in frames or foundation. Bee disease insurance is therefore less attractive to them.

The other hazard not unknown in beekeeping is back injury. A Warré box completely full of honey can weigh 21 kg (46lb). Clearly caution and correct lifting technique are called for. Occasionally, when inserting a third box, I lift two Warré boxes complete with bees and almost full of comb. However, this is generally early- to mid-season before a significant weight of honey, the densest material in the hive, has built up.

1.6 Recommended reading

Apart from my own book and Warré's, both mentioned in the Preface, I would like to recommend a few books by other authors that should prove especially useful to a novice natural beekeeper. Michael Weiler's *Bees and Honey from Flower to Jar* (Floris Books, 2006) is an inexpensive book covering bee anatomy, biology and behaviour. It really addresses the phenomenon of *Der Bien*, the essence of the bee. Günther Hauk's *Towards Saving the Honeybee* (Biodynamic Farming and Gardening Association,, 2002) presents several principles and practices of natural beekeeping particularly from a biodynamic perspective. For its superb colour photography and pictorial description of bee biology and behaviour, Jürgen Tautz's *The Buzz About Bees – The Biology of a Superorganism* (Springer 2008) is a must read. It is expensive, but your public library will no doubt obtain it for you. A little book that you will find invaluable to have handy is Heinrich Storch's *At the Hive Entrance – Observation Handbook: How to Know What Happens Inside the Hive by Observation of the Outside*. (European Apicultural Editions, 1985). For a more academic/scientific account of bee biology and behaviour I suggest Mark Winston's *The Biology of the Honey Bee* (Harvard University Press, 1987). For a thoroughly entertaining read, I recommend any bee book by Tom Seeley and in particular his *Honeybee Democracy* (Princeton University Press, 2010) which describes his research on how bees find a new home. Anyone working with natural swarming will very likely want to know what is happening in space and time between their hives and their bait hives.

To supplement the books, there are several web sites devoted to Warré beekeeping, the oldest in English of which is www.warre.biobees.com, the web space for which is kindly provided by Phil Chandler who is a UK pioneer of natural beekeeping based on the horizontal top-bar hive. My own website www.bee-friendly.co.uk includes Warré beekeeping material that is not available on the biobees.com site as well as my articles and links to my work on bee colony relocation. The site of the 'warrebeekeeping' e-group, mentioned above, has several years worth of searchable messages, many now compiled into a list of frequently asked questions. Gill Sentinella's film on DVD entitled *The Honeybee* (2009) gives a vivid account of the annual cycle with superb photography.

2   Getting your Warré hive

Since Warré's book appeared in English translation, at least a couple of dozen manufacturers/ suppliers have sprung up round the world. Particularly encouraging was the decision of E. H. Thorne (Beehives) Ltd., Britain's biggest beekeeping supplier, to bring out a 4-box Warré in English cedar in 2011. This marked a certain recognition of the hive's merits, at least as a commercial proposition. In Appendix 2, I list the Warré suppliers known to me at the time of writing this book.

Warré himself, however, envisaged 'The People' themselves making his hive as it is so simple in design. Therefore here I present full instructions for its construction, including conversions from Warré's metric measurements to imperial. In addition, I include adaptation of the box to contain a window at the back and modified floors.

2.1 Materials

The choice of wood species is largely up to the beekeeper, depending on local resources and how much is to be spent on the hive. Many of my Warrés are made from recycled wood. However, this is not universally accessible. Softwood is preferable for its lower thermal conductivity and therefore greater insulating power. Of the commonly available softwoods I would recommend, in descending order of preference: cedar, larch, pine, and spruce – all ideally from sustainably managed local forestry. Western red cedar is the traditional wood for beehives in the UK as it is lightweight and durable without any outer protection. Many of my newer boxes are made from larch, but I still use recycled wood for all other parts of the hive.

If you are limited to buying planks already cut to size, say in pine, you may wish to adapt the hive dimensions a little in order to suit what is available. For example, 1" x 8" timber/lumber when planed all round is ¾" x 7¼". If you use it planed you would lose about 1" depth on your boxes, therefore have more brood nest interruptions than otherwise, although correspondingly lighter boxes to lift. The bees won't mind. But you could use 1" x 10" (¾" x 9¼") which would mean boxes an inch deeper, and correspondingly heavier when filled with honey. If a woodworker would rip saw an inch from the width for you, you would have boxes almost exactly as specified by Warré. Alternatively you could use 1" x 8" with a sawn finish, i.e. no wood wasted in planing. Indeed, Warré stated that there is no need to plane the wood. Bees do not have hollow trees planed for them! There are two advantages. Your boxes will be thicker, therefore better insulated, and the box depth would be only about a quarter of an inch shallower than Warré's design. Your bees would not complain.

Yet, as far as possible, try to maintain the box internal dimensions with a footprint of 300 x 300 mm (11¹³/₁₆" x 11¹³/₁₆") so that you do not run into difficulties with the fitting of top-bars and making spaces between them.

Some Warré beekeepers use exterior grade plywood and even marine plywood for parts of the hive. It is not known to what extent substances used in the adhesives in the ply would diffuse out, but it is certainly a factor to consider. It should of course be no problem if ply is used in the roof, and even the quilt. To simplify construction, some do use ply on the floor, even though it is in direct contact with the bees. Three other factors to consider with plywood, compared with solid wood, are its higher cost, lower vapour permeability, and higher thermal conductivity, i.e. higher heat loss.

Some natural beekeepers like to avoid all metal or plastics in their hives. Ferrous metals might interfere with the bee's perception of the earth's magnetic field. Whilst it is true that bees can sense this field, it is not known if proximity to steel causes any interference. If you want to be on the safe side, you could use copper nails or even finger joint and glue your box corners. To avoid glue chemicals coming into contact with the bees, it could be applied to the outer parts of the joints. Any inner gaps will be filled by the bees with propolis, a brownish sticky substance which is their natural adhesive, mastic, and antiseptic. I have made a few boxes with finger joints. It is time consuming and requires great accuracy, and a greater degree of woodworking skill. In most of my boxes I use galvanised nails.

Fig. 2.1 Warré box with finger jointed corner

Use of plastics in hives is frowned on because they are believed to release substances of questionable food safety that are loosely trapped during the manufacture. It is easy to do without plastics most of the time. However, plastic containers are just too convenient not to use them occasionally as feeders. If plastic is a concern, ceramic, glass, food grade stainless steel, or enamelware are other options.

Wood preservatives are best avoided. I finish the outside of my boxes with two coats of hot raw linseed oil, allowing 24 hours between coats. Some mix beeswax into the oil, e.g. 25 g wax per 500 ml oil (1oz/pint).

Observe fire precautions regarding hot oil and the storage of rags/brushes used to apply it. Some prefer tung oil. I use exterior grade paint on the roofs. If you can obtain ecological paint, so much the better. I use nothing on the floors, quilts or stands. Even the stand, which is the component most prone to rot, is frequented by bees. However, I regard my stands, all made from recycled wood, as disposable. More durable stands could comprise two concrete blocks resting on a paving slab, preferably all three items recycled.

Other materials used will be discussed as they arise. The plans shown below are based as closely as possible on those in the 12th edition of *Beekeeping for All* published by Warré (1948).

In § 2.2.9 are plans for constructing a box with a window according to Frères' and Guillaume's modification of Warré's hive.[5] I include this variant of the box because many beekeepers, especially beginners, find it helpful for monitoring the progress of colony development. But adding windows increases complexity, expense, and consumption of resources. They need care in design and use to avoid compromising hive thermal performance, e.g. due to warping, or disturbing the winter cluster.

The plans shown here are based on boxes of 20 mm (¾") thick wood, the minimum that Warré regarded as sufficient. However, he recommended 24 mm (1") for improved

---

[5] Frères, J-M. & Guillaume, J-C (2013) *L'Apiculture Écologique de A à Z*. Marco Pietteur, Belgium

rigidity. Thicknesses of 38 (1½") and 50 mm (2") are used in colder climates. Such boxes are heavy. Any change to the box wall thickness should ideally retain the internal measurements of 300 x 300 x 210 mm (11^{13}/$_{16}$ x 11^{13}/$_{16}$ x 8¼"). This will require all other components of the hive apart from the legs to be re-sized accordingly.

2.2 Construction

2.2.1 The hive

The hive comprises between two and five (rarely six) stacked boxes each fitted with handles and eight top-bars, the bottom box resting on a simple floor (bottom board) that is notched to give an entrance under the box rim. The top-bars of the top box, i.e. box 1,[6] are covered with a sized top-bar cover cloth on which rests a wooden-framed quilt (*coussin* in the original). On the quilt rests the eaves- and gable-vented roof whose bottom rim must come about 20 mm (¾") below the joint between the quilt and the top box, in order to keep out rain. There is no direct passage of air from the top box to the vents under the roof. The quilt is filled with plant-derived insulating material. Some use wool or vermiculite. As well as the top-bar cover cloth, there is a cloth fixed to the bottom of the quilt to retain its contents.

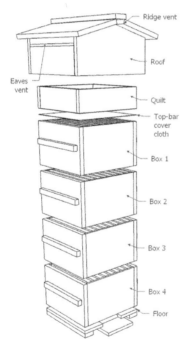

Fig. 2.2 The People's Hive exploded   Fig. 2.3 The People's Hive

2.2.2 The hive-body boxes and top-bars

The basic box is butt-jointed at the corners. Nailing with seven galvanised nails 65 x 2.65 mm (2½ x ⅛") makes a strong joint. Four screws of the same length also suffice. Glue

---

[6] Warré always numbered his boxes from the top downwards. So do I, on the grounds that the first box to be populated is the top one, namely box 1.

is unnecessary. Obviously half-joints or finger joints make a stronger box, although this increases the complexity. Any cracks are filled by the bees. However, it helps if, from the outset, the boxes sit on each other without rocking too much. Thin gaps are anyway propolised by the bees. Use a square throughout assembly. If in doubt, knock nails in only partially at first. To compensate for any curvature in planks, guided by the rings showing on the end grain, face the outside of the tree to the inside of the joint that is being nailed.

Warré does not specify the size of the handles, but 300 x 20 x 20 mm (1' x ¾" x ¾") bars, fixed with three nails and glued, have been found to work well in practice, including when lifting the entire hive with the bottom box handles. I slope the upper surfaces of the bars to shed rainwater.

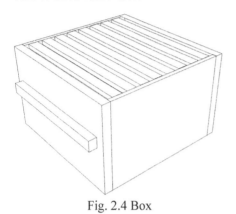

Fig. 2.4 Box

The spacing between all bars and between the end bars and the walls is 12 mm (½"). Cutting the rebates to support the bars can pose a problem for the do-it-yourself woodworker. Normally rebates would be cut with a table saw, a router, a rebating plane, or even with a hand saw provided that a guide is affixed to the wood. However, as their depth is less than half an inch (10 mm) they can also be cut with a woodworking knife (e.g. Stanley) using a steel ruler as a guide. Make several cuts of increasing depth down through the wood. Avoiding rebates and instead resting the top-bars on battens, an alternative that Warré suggests, is inadvisable as it makes later comb removal, for example for official inspection, extremely difficult.

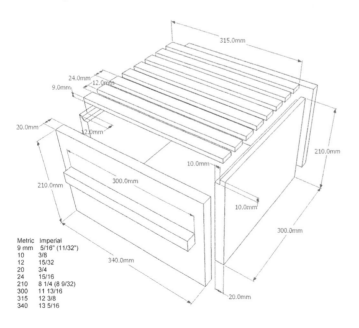

Metric	Imperial
9 mm	5/16" (11/32")
10	3/8
12	15/32
20	3/4
24	15/16
210	8 1/4 (8 9/32)
300	11 13/16
315	12 3/8
340	13 5/16

Fig. 2.5 Box exploded. (The imperial measurements shown in brackets are closer to Warré's original specification.)

Jean-François Dardenne has suggested putting rebates at the top and bottom of the same walls of the box. The idea behind this is that it prevents the bottom rim of a box from being propolised to the ends of the top-bars of the box below it. I use such double rebates on all my boxes.

Warré specifies a top-bar length of 315 mm (12³/₈"). If it is extended to approach 320 mm (12⁹/₁₆"), then it better spans the rebate and leaves no end cavities for pests, such as small hive beetle, to hide in. Notice that the top-bars are slightly shallower than the rebate's depth. This is Warré's detail ensuring that boxes sit flat and no top-bar projects above the top rim of the box, thereby making box removal difficult.

Top-bars can be of any wood that will stand the flexing of 2 kg (4lb) comb. I use offcuts or scraps of pine, larch, oak, cedar, mahogany, etc. A table saw greatly facilitates their cutting, but be warned that beekeepers are known to have nasty accidents when cutting top-bars. Top-bars should be left rough-sawn underneath to maximise comb binding through providing a key for the wax, but planed on top to minimise binding to the combs above. If your bars are already planed on both sides, place a handsaw in a vice, and, wearing garden gloves, rub the undersides of the bars back and forth over the teeth to provide a rough surface. The top surface only may be coated with raw linseed oil which weakens adhesion by any combs above. Two coats are suggested, allowing at least 24 hours in between. Take care not to get the oil on the sides or undersides of the bars. This is important for ease of removal of boxes. It must be done several days in advance of use in order to allow the oil to 'dry'. Do not use boiled linseed oil as it sometimes contains chemical drying accelerators. If you have forgotten to oil your bars in advance, a last-minute coating of Vaseline (petroleum jelly from the pharmacist) can be used. It should be coated extremely thinly by working it into the grain, and the excess scraped or rubbed off. Although Warré also recommends its use, we should consider that it is less immediately plant-derived than linseed (flax) oil, and likewise less edible. However, in the very small amounts used, it can be considered safe for contact with honeycombs.

In extremely cold climates, with snow on the ground for weeks or months, providing a small top entrance has proved beneficial. This may seem to go against Warré's plan of retaining heat at the top of the hive and maintaining a seal there most of the time. However, if the bees cannot otherwise get out to defecate, a serious health issue could arise. The entrance could be a slot 6 mm (¼") deep spanning the gap between two middle top-bars at the front of the top box, or a 10 mm (³/₈") hole in the front wall of each box and plugged when unused. The top-bar cloth, quilt and roof are installed just the same. When replacing the roof, most of the gap between the roof and the box is positioned at the front, thus leaving a sheltered exit a little above the bottom rim of the roof. Other ways of arranging this top entrance are conceivable, but they mostly involve dispensing with Warré's quilt. The top entrance may of course be closed at any time by inserting a plug of wood of the same size as the slot.

2.2.3 Waxing, positioning and fixing top-bars

Each top-bar has a bead or strip of wax applied to its centre underneath to provide a guide for the bees in positioning comb. Without comb guides (starter strips), the bees would build combs under the bars at their chosen angle and spacing, making comb removal on top-bars impossible. Beginners will be able to get beeswax from a local hardware store, woodworking suppliers, candle shops or other beekeepers. As it is quite likely to be contaminated with pesticides including acaricides (miticides) you may wish to get it from a beekeeper keeping his bees organically. However, as relatively little is used, compared with the amounts used in foundation in frame hives, you may fairly safely use ordinary beeswax if you cannot get wax from organic beekeeping. Some like to have a groove cut down the middle of their top-bars to provide a key for the wax guide. This modification, suggested by Frères and Guillaume, is unnecessary.

Fig. 2.6 Waxing tools

To wax the top-bars you need a heat source, a pan with water to act as a bain-marie, a used food can to contain the wax, an old teaspoon, and a template. The template is a piece of wood the length of a top-bar and 15 x 24 mm (5/8" x 1") in section. On one of its wider sides two pins or small nails are inserted near the ends and half the width of a top-bar from one side. Run a little cold water into your sink and put the template to soak in it. Melt the wax in the bain-marie. When the wax is melted, warm the spoon in it. Shake any excess water off the template, and, with the top-bar's underside uppermost, place the template on it with the pins resting against the top-bar's edge, so as to line up the edge of the template with the middle of a top-bar (Fig. 2.7). The centre of the top-bar must not be wetted with water or the wax will not stick. Hold the two pieces of wood to form a 'V'-shaped trough sloping slightly to one end, with the template providing the surface on which most of the wax will be deposited. Spoon the wax into the trough, especially against the template, allowing the wax to run down the slope along the bar. In

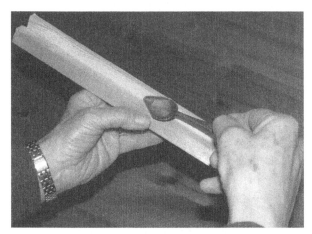

Fig. 2.7 Waxing a top-bar

this way, spooning more wax as required, deposit on the template a thin film of wax up to about 2-5 mm (1/16 - 3/16") at its base on the top-bar. The height of the film when viewed from the side may be anything between about 5 mm (1/4") and the width of the template.

When all the wax has set, usually within a few seconds, gently remove the template, leaving behind an undulating wall of wax aligned with the middle of the top-bar. Scrape away any wax that is within 10 mm (3/8") of the ends so that the bars will sit properly in the box rebates. Repeat the process with all eight top bars for each box to be prepared.

For large numbers of top-bars, it has been found convenient to keep the melted

Fig 2.8 Top-bar with wax starter-strip

wax in a squeezable washing-up liquid container, which stands in the bain-marie when not in use. The wax is then simply squeezed from the nozzle. A smaller-scale alternative to this is a piece of copper water pipe, constricted at one end and used as a pipette. The constricted end is dipped into the wax and the index finger placed over the other end. Wax is run in a controlled manner from the pipette onto the top-bar according to how much the index finger is released.

Warré advises pinning the top-bars in position at 36 mm (1$^7/_{16}$") centres with headless glazing pins. The suggested actual imperial sizes and spacing are given in Fig. 2.5. I follow Warré's advice except that I use hive frame pins (UK: 20 x 1 mm japanned, ¾"). I drive them in about 15 mm ($^5/_8$"), snip the heads off (protect eyes!) and hammer them flush. Do not use nails as it makes top-bar removal difficult. To achieve the correct spacing, I use

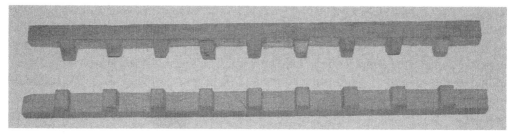

Fig. 2.9 Top-bar spacing templates

two top-bar spacing templates as shown in Fig. 2.9. A card or paper template is easier to make. Some even space the bars by eye!

I fix top-bars so as to have the whole box ready to install on reaching the apiary. It also avoids the risk of combs in a lower box lifting when removing a box above. However, nailing top-bars makes comb removal at the rare inspections more difficult. Furthermore, you may wish to defer fixing top-bars because one method for hiving a package of bees (§ 6.2.1.1) involves using two boxes without fixed top-bars, and only later positioning them.

Fig 2.10 Removable top-bars with locating pins. Photo: Larry Garrett

Some fix their bars less firmly in place using melted beeswax. Others cut slots or drill small holes in the bar ends that fit over headless nails exactly positioned in the rebates. Castellated metal spacers are also used. These are already available to fit Warré boxes. Matchsticks pushed into holes drilled in the rebates is another way of keeping the bar spacing accurate. Rarely are bars not secured at all, although after a short time the bees see to it that the bars are firmly propolised in place. The risk of not fixing the bars in advance is that if the hive is disturbed shortly after installing a new box, the bars may shift out of position. It is a good idea to weigh all boxes complete with top-bars before use so these weights can be taken into account when hefting later (§ 9.3).

2.2.4 The top-bar cover cloth and its sizing

A suitable material for both the top-bar cover cloth and the quilt contents retention cloth is hessian (burlap), a coarsely woven fabric made from jute that is traditionally used in sacking, e.g. peanut sacks (pet shops) or coffee bean sacks, washed if necessary. In any case, use food grade sacks, as others may be treated. Hessian from fabric, decorating or upholstery shops should also be treatment-free. Heavy-duty canvas made of cotton, e.g. traditional sailcloth, is a satisfactory alternative and it does not need sizing. Canvas might have a slight disadvantage in that there are no small holes in it for the bees to propolise or unpropolise in order to control ventilation, a process which Warré as well as Frères and Guillaume claim to have observed. However, I have not yet observed it on my hessian top-bar cloths.

A few Warré beekeepers use the woven polypropylene sheeting used in various applications, such as ground covering, builders' aggregate bags, sacks, etc. Although this material is 'breathable', i.e. fairly well vapour permeable due to its woven nature, it means introducing plastic of questionable food compatibility into the hive.

Occasionally, bees chew small holes in sacking top-bar cloths, even when they have been sized properly as Warré advises. I have noticed this rarely and have had few reports of it from others. In only one case did the bees chew through both cloths, and quilt insulation was seen being dragged out of the entrance! If your bees seem particularly inclined to chew the top-bar cloth, you may need to resort to the aforementioned polypropylene sheeting, or, better still, heavy-duty plastic fly screen available from hardware shops. Indeed, it is precisely this and not sacking which Jean-Claude Guillaume recommends in his book, which is influential amongst Warré beekeepers in France.[7]

Fig. 2.11 Sizing a top-bar cloth

Warré advises cutting the cloth to fit the hive *after* the cloth has been sized. I have found this to be unnecessary. Flour paste is used for sizing the cloth to prevent the bees from chewing holes in it. Stir 125 g (1 cup, 4½ oz.) wholemeal flour, preferably of rye, and 10 g (1 tablespoon, ½ oz) clothing/laundry starch (e.g. maize starch) into a pan containing 1.5 litres (6⅓ cups, 3½lb, approx. 3 US liquid pints) of cold water until thoroughly dispersed without lumps. Bring to the boil while stirring. When the liquid thickens like gravy, allow to cool. Place the cloths on a non-absorbent surface. Apply the paste to both sides with a brush or the fingers, working paste into the fibres. I find that this amount just does 18 cloths. Place the wet cloths on a non-absorbent surface to dry. They dry overnight indoors, and probably within a few hours in the sun.

When dry, the cloths are normally so rigid from the sizing that they can be held horizontally

[7] Frères, J-M. & Guillaume, J-C (2013) *L'Apiculture Écologique de A à Z.* Marco Pietteur, Belgium

when slightly curved. Although the sizing is invisible when dry, it has been said that flour paste is not a natural thing to put in contact with a bee colony. I have suggested, though not tried, the alternative of soaking the top-bar cloth in beeswax.

2.2.5. The quilt

Warré's quilt was 335 mm (13³/₁₆") square as he brought the quilt cloth up its sides for fixing and needed the extra space created inside the roof for the quilt contents retention cloth. I make my quilts the same size as the boxes and fix the cloth to its underside with staples or tacks. Larry Garrett (Indiana) suggests that bringing the hessian/burlap out at the sides of the quilt box performs a ventilation function, and that the weft and warp of the hessian sacking (burlap) weave provide a capillary wick from any area of the cloth to each of the hive's four exterior sides. The cloth needs no special preparation.

I sometimes replace the fixed quilt cloth with a bag loosely filled with quilt filling. The bag may then be lifted to install a small contact feeder in the quilt frame, and replaced over and round it to retain the heat.

The thickness of the walls of the quilt could be as little as 10 mm (³/₈") as it supports only the roof. The plant-derived insulation it contains may be chopped straw, wood shavings, dried leaves, shredded paper, etc. Two Warré users in the USA have found that if there is a problem with ants in the apiary, eastern red cedar shavings keep them out of the quilt.

In the extremely cold parts of Canada the Warré-type quilt has been rejected as inappropriate. The concern is that as the temperature stays below -20°C (-4°F) for many weeks, sometimes reaching -35°C (-31°F), condensation of water from the consumption and metabolism of honey might form and freeze in the quilt and then drip onto the cluster in the spring thaw. John Moerschbacher, who recently moved from Alberta to British Columbia, has therefore substituted for the quilt a wooden feeder with central access holes to the tank. Any water vapour entering it condenses under the roof and collects there, providing a water supply for the bees if they need it for diluting honey during the early spring build-up.

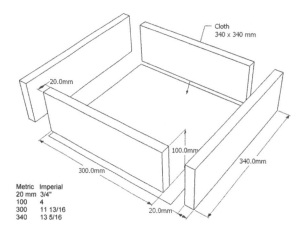

Fig. 2.12 Quilt

Metric	Imperial
20	3/4
100	4
340	13 5/16

Fig. 2.13 Quilt exploded

Metric	Imperial
20 mm	3/4"
100	4
300	11 13/16
340	13 5/16

AFTER FIXING # HESSIAN - FIX A 10x20 STRIP TO EDGE OF BOX - THIS WILL KEEP HESSIAN AWAY FROM BARS

2.2.6 The roof

This is the more complicated of the two roof designs that Warré presents in *Beekeeping for All*. It has a ventilated cavity immediately under the upper surfaces of the roof. This is to dissipate solar heat. Inside is a cover board (or 'mouse board') which rests on the quilt. Thus there is normally no passage of air from the quilt to the ventilated roof cavity. This board could be made of a row of strips of offcuts up to 13 mm (½") thick. Some Warré users have occasionally seen condensation under the mouse board. Strips of wood allow water vapour to diffuse through the cracks. If using a solid board, consider drilling a couple of dozen 3 mm (¹/₈") holes in it.

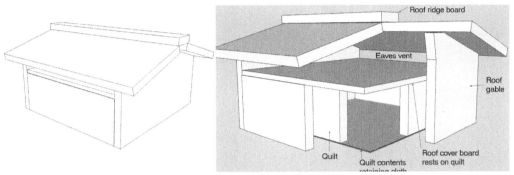

Fig. 2.14 Roof   Fig. 2.15 Cut-away view of roof and quilt

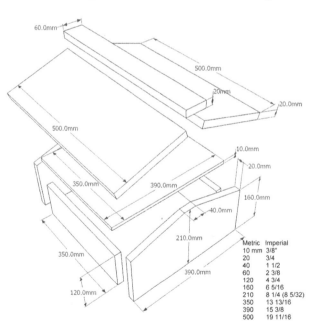

Fig. 2.16 Roof exploded

Metric	Imperial
10 mm	3/8"
20	3/4
40	1 1/2
60	2 3/8
120	4 3/4
160	6 5/16
210	8 1/4 (8 5/32)
350	13 13/16
390	15 3/8
500	19 11/16

The example given here uses 20 mm (¾") wood for the outer structure. This is based on Warré's plans. Wood of 13 mm (½") thickness would be perfectly adequate and result in a roof lighter to lift. There is 10 mm (³/₈") play between the roof inner walls and the quilt outer walls to ease placement and removal of the roof. Roof imperial measurements could be further rounded provided about half-an-inch play (³/₁₆ to ¼" all round) is maintained and its bottom rim comes at least ¾" below the quilt base.

I put a brick or a stone on the roofs of my hives.

If thin wood, more likely to warp, is used for the inclined boards, they may be nailed to the ridge board from their undersides by supporting the three pieces upside down, with pieces of wood suitably positioned to hold the boards at roughly the correct angle. This unit of three boards is then nailed to the gable ends. Nailing the sloping boards to the ridge board helps form a better seal against driving rain.

Fig. 2.15 shows a cut-away view of the quilt in the roof. Note that the lower rim of the roof projects below the junction in which reside the top-bar cloth and the cloth fixed to the bottom of the quilt. This prevents rainwater from running into that joint and wicking into the top-bar cloth and quilt.

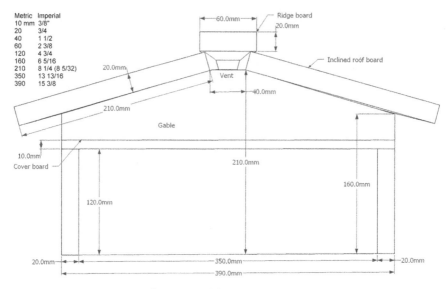

Fig. 2.17 Gable elevation of roof

2.2.7 The floor (bottom board)

Warré recommended 15-20 mm ($^5/_8$-$^3/_4$") for the thickness of the floor. He gave no guidance on the thickness of the battens underneath it.

The notch in the floor serves as the entrance and is 40 mm (1½") deep (front to

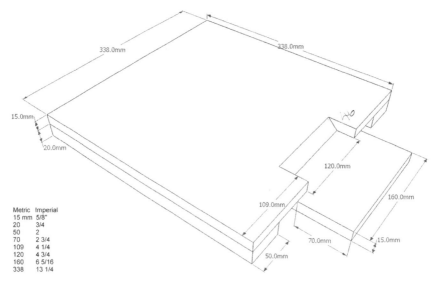

Fig. 2.18 Floor

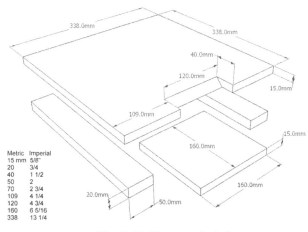

Metric	Imperial
15 mm	5/8"
20	3/4
40	1 1/2
50	2
70	2 3/4
109	4 1/4
120	4 3/4
160	6 5/16
338	13 1/4

Fig. 2.19: Floor exploded

back) for hive walls of 20 mm (¾") thickness. If required, the alighting board can extend right to the back of the floor to give the latter added rigidity. I like to plane a slope on my alighting boards to help rainwater drain off.

Note that the floor is 2 mm ($^1/_{16}$") narrower in both horizontal directions compared with a box. Warré suggested this to promote drainage of rainwater.

2.2.8 The legs

The leg has a wide foot attached to prevent the hive from sinking into the ground and possibly toppling. Furthermore it gives a 20 or 30 mm (¾ or 1¼") projection outside the footprint of the floor, thus greatly increasing stability. A few Warré hives have been toppled through having legs/feet or a stand that are of an unsuitable alternative design. If a separate stand is made of wood or concrete blocks, it is worth ensuring that its outer corners are a little outside the hive's footprint.

The leg places the Warré hive entrance relatively close to the ground compared with most hives. However, Warré regarded a low entrance as important (pages 46-48, *Beekeeping for All*).

In Fig. 2.20, the right hand pillar of the leg has the underside of the floor resting on it. The angled left hand pillar is nailed from both sides to the side of the floor with a total of four nails. The difference in height of the right and left pillars is 25 mm (1"), which will work for the floor corner

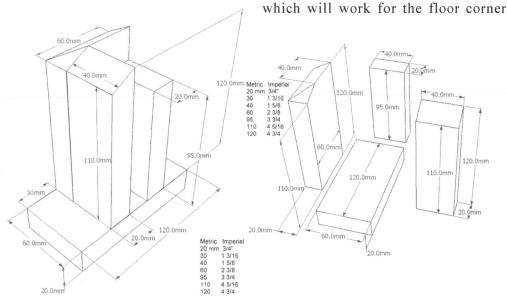

Metric	Imperial
20 mm	3/4"
30	1 3/16
40	1 5/8
60	2 3/8
95	3 3/4
110	4 5/16
120	4 3/4

Fig. 2.20 Leg unit

Fig. 2.21 Leg unit exploded

overall thickness of 35 mm ($1^3/_8$") shown above. If the floor corner is less than 25 mm (1") thick, then the right pillar should be lengthened accordingly. The feet form a pinwheel (Catherine-wheel) pattern which maximises stability.

2.2.9 Box with window and shutter

Warré boxes with windows appeal especially to beginners, but once one is used to the dynamics of the Warré and to other ways of monitoring progress, windows seem less necessary. This modification of Warré's box is based on the design by Frères and Guillaume.[8] They base some of their management decisions on what is visible through the windows at the back of the hive. For example, if they see that there is almost no honey left at the top of the top box, the bees are

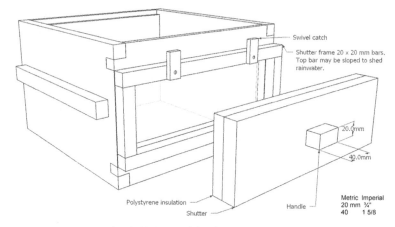

Fig. 2.22 Box with window and shutter

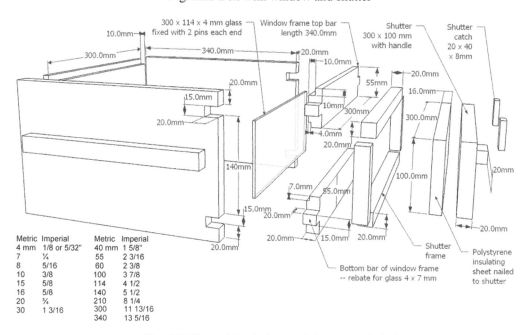

Fig. 2.23 Box with window and shutter exploded

[8] Frères, J-M. & Guillaume, J-C (2013) *L'Apiculture Écologique de A à Z*. Marco Pietteur, Belgium

possibly nearing starvation. The window helps in monitoring comb growth. However, once a box is full there is generally little to observe through windows placed at the back, as is normally the case. Occasionally one sees the queen there and on one occasion I was able to watch some worker bees help a recently emerged 'princess' to sting her sisters to death in their cells.

Instead of windows, which add to the cost and complexity of the hive, one might wish to drill holes of about 25-50 mm (1-2") near the base of the front of each box. These may be closed with a wooden plug or sliding cover when not in use. Some Warré beekeepers close them with a rotateable plastic entrance disc which also includes a queen excluder and a ventilation grille (available from beekeeping suppliers). These holes provide additional entrances which in some localities are found to be a boon. However, bear in mind that each extra entrance needs an additional set of guard bees.

The window shutter's insulation should be flush against the glass with the smallest gap all round compatible with easy removal. The depth of the cavity exactly matches the depth of the shutter plus insulation. To minimise jamming of the wooden part of the shutter in wet conditions, it may be made a millimetre or more smaller in each direction. Refrain from uncovering windows in winter.

In this version, the window frame bars are jointed to the box walls with mortise and tenon joints. Alternatively, butt joints with three nails or two screws at each joint should give satisfactory service.

2.3 Ekes, modified floors, sumps, and feeders

An eke, in its shallower form sometimes called a 'shim', is a hive body of any depth, from perhaps a couple of inches to the full height of a standard box. It can be used above or below the brood nest, for inserting, for example, feed dishes/jars or candy and for delivering varroa treatments. Commonly the Warré beekeeper will use as an eke an empty quilt frame with no retaining cloth.

Some Warré beekeepers modify their floors in various ways. A wire mesh screen for monitoring varroa mites can be incorporated. As I do not treat for varroa, I have long since given up counting mites, but those who wish to practise integrated pest management may need a mesh floor. A problem with mesh floors is they create corners for wax moth larvae that the bees cannot get at. In regions where small hive beetles are troublesome, a beetle trap may be incorporated into the floor.

Perhaps a more versatile modification is a floor which includes several features in a single unit: 1. rear opening; 2. removable mesh and tray for monitoring varroa; 3. wide enough to admit a camera or a light and a mirror for monitoring colony progress; 4. enough space for inserting a dish of feed; 5. closable at the rear so that intruders cannot enter. This again runs the risk of wax moth build-up and certainly requires debris to be removed at regular intervals, because in such a design the bees do not nor-

Fig. 2.24 Mesh floor with drawer

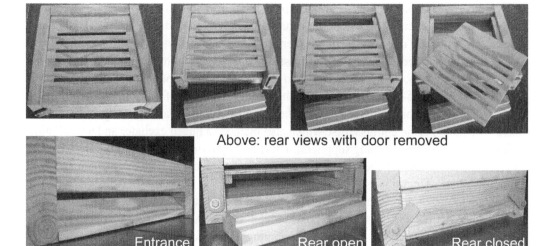

Above: rear views with door removed

Entrance   Rear open   Rear closed

Fig. 2.25 Floor with rear access. Photos: Larry Garrett

mally have access to the lower space, whereas floors of the Warré type are generally kept clear by the bees.

Instead of a Warré floor on a stand, some like to seat the hive on a box about the same depth as a hive-body box, which sits on a concrete slab, thus forming a sump. A slot is cut in the upper rim of the bottom box to serve as an entrance. There may or may not be an alighting board. The idea behind it is the creation of an internal environment similar to the bottom of a hollow tree cavity. Also, mites drop well clear of entrance traffic. A sump could have rear access for cleaning, feeding or viewing colony growth.

I have never found the need to have modified floors in permanent use. However, I can see that they could reduce the amount of hive lifting needed. When the hive gets too heavy for me to lift alone I use a Gatineau lift (§ 4.4).

As you will want to be planning ahead, I should mention feeders at this stage. I deal with them in more detail in § 11.1. Warré describes a low capacity spring feeder and a large capacity autumn feeder. Both are made of wood, albeit with the addition of a suitable sealant. It would be advisable before you get any bees to decide on the best options for you as regards feeding. The hive is very flexible in this respect, and you may be able to do entirely without the feeders that Warré describes. However, you may wish to check the options and, if necessary, order equipment in advance.

2.4 Mouse guards and robber guards

Warré suggested making a special entrance piece cut from a food can to serve as a winter entrance, mouse guard, and robber guard, depending on how it is installed (page 54, *Beekeeping for All*). A mouse guard needs to be narrow enough not to admit the skull of the mouse. Seven and a half millimetres ($^5/_{16}$") is generally considered adequate. My mouse guards comprise a straight piece of rectangular 1.5 mm ($^1/_{16}$") thick metal (7.5 x 140 mm; $^5/_{16}$" x 5½") with a small hole drilled close to each end. To fit it, a 7.5 mm ($^5/_{16}$") thick spacer block is rested on the alighting board and the guard, resting on it, pinned with drawing pins (thumb tacks) to either side of the entrance through the holes provided (see Fig. 12.1, page 82).

All sorts of items can be deployed as robber guards if the emergency arises, pebbles, blocks of wood, handful of grass, etc. However, if a robber guard has to be fitted in hot weather it is probably best to cut it from bee- and wasp-proof metal mesh (gaps less than 3 mm; $^1/_8$") so that ventilation is not seriously compromised. With mesh, the entrance can be as small as one bee wide.

2.5 Stands and stability

If you decide not to fit some kind of leg to your hive floor you will need a stand. As a couple of concrete blocks suffice I will not go into the details of the construction of my stands which place all four legs outside the corners of the floor (Fig. 2.27). This feature also provides a resting place each side of the hive for the feet of my hive lift. I rest my stands on recycled concrete slabs. Warré comments 'I have seen such stands made from a framework whose wood would have been sufficient to make a double-walled hive.' This criticism is offset by using recycled wood, scrap or offcuts.

Fig. 2.26 (left) Simple hive stand using concrete blocks.
Photo: Trevor Ray
Fig. 2.27 (above) A very stable hive stand suitable for the
Gatineau lift (§ 4.4)

Warré preferred his floors to be only about 100-150 mm (4-6") from the ground. Mine are 300 mm (1 ft) and I have never found the extra height to be a problem. I have never had any of my Warré hives blown over, despite some being five boxes high and receiving gusts of up to 80 mph. In 2012, a Warré hive in Pennsylvania, 10 miles from the eye of hurricane Sandy, remained standing even though the neighbour's garage roof was ripped off, whereas Langstroths in the vicinity did not fare so well. However, I have heard of three incidents of Warré hives blowing over, so it is worth taking some care regarding stability and wind protection. In areas often exposed to high winds, some use hive straps to secure the hive to the stand or ground.

Fig. 2.28 Hive stands in Switzerland
Photo: Uli Schläpfer

If a hive rises to six or more boxes, you may wish to secure it to a stake driven into the ground.

Mature colonies in a Warré hive are astonishingly robust. A large ash tree fell on my apiary in a storm in January 2012. Two hives were struck by it, one a frame hive, the other a Warré. Both were knocked off their stands and inverted. Frames were spilled onto the ground, but the Warré remained intact with all combs still firmly fixed in place and the two boxes stuck together with propolis. At the time of writing, a year later, the colony in it is still alive.

2.6 Frames and semi-frames

Some beekeepers like the narrow format of the Warré hive, yet require it to have frames or semi-frames instead of merely top-bars, perhaps because the possibility of rapid comb removal is a requirement in their state/country. I have even heard it argued that framing the comb makes a hive more bee-friendly. However, that may apply only to those unwilling to take the extra care required when handling comb on top-bars.

A frame version of the hive, slightly wider to accommodate the extra wood, is described in the 5th edition of Warré's book.[9] Ickowicz (France) sells a frame version of the hive. Semi-frames are a compromise between top-bars and frames. The side-bars are about half the length of frame side bars and there is no bottom rail. I do not use the frame version of the People's Hive, although I do occasionally use a few frames crudely constructed with scant regard for so-called 'bee space' and fitting my standard Warré boxes. These are for hiving colonies that I have rescued from buildings. Cut-out comb is secured in the frames with rubber bands. However, I do not discuss frame beekeeping in this book, which is intended for those requiring more natural comb.

Roger Delon, who died in 2007, developed a kind of compromise frame comprising a top-bar attached to a 'U' shape of stainless steel wire. Its compromise lay in the fact that combs could be built past the wire to the hive walls, thus maintaining the retention of nest scent and heat, but were more easily removable without breaking.

[9] Warré, E. (1923) *L'Apiculture Pour Tous*, 5th edition, Tours. http://warre.biobees.com/warre_5th_edition.pdf. Pages 60-71 translated by David Heaf: http://warre.biobees.com/warre_5ed_60-71.pdf.. See also JPEG images of scans of the pages of the original book at http://ruche.populaire.free.fr/apiculture_pour_tous_5eme_edition/.

3 Siting your hive

First familiarise yourself with any bylaws or ordinances that apply to siting hives. Then consider discussing your plans with family and neighbours. I have had some apiary neighbours who really love having my bees near them, and others who are terrified of them.

It should go without saying that there needs to be sufficient forage in the locality. This generally applies in lowland Britain, but in 2012 it emerged that, due to a surge in hobby beekeeping in London and New York, hive numbers reached levels that could not be sustained by the urban forage available. By October 2012 there were over 3,000 hives in central London. Therefore, prospective beekeepers in any locality would be wise to check local colony densities and potentially competing apiaries.

Common sites include gardens, back yards, city rooftops, allotments, field margins (fenced against cattle) and wasteland. Walls, fences, hedges, hurdles and/or screening nets (windbreak) may help guide flight traffic in the desired direction and protect the hive from winds. The flight path near the hive entrance should not cross thoroughfares or any place where people frequently pass. Not siting too close to places frequented by people will help you too, as you may want to open the hive on a sunny day, just when everyone else is outdoors. Furthermore, if the hive is out of sight, there is less chance of it being targeted by vandals or thieves.

Here, in wild, wet, windy, west Wales, especially with the long sequence of cloudy summers we have been having, I like my hives sited in the sunniest positions available. However, in warmer climates, direct sun, especially in the afternoon, can be

Fig. 3.1 Warré apiary in field corner

seriously damaging to the colony. In such locations, arrange for full or partial shade. Remember that in tree hollows, nests are shielded by a thick wall and shaded by leaf cover. But don't go to the other extreme: a damp, dark, dingy corner is not good for colony health. There's an old saying amongst UK beekeepers: bees in a wood never do any good. I have never tested it out. Good air and water drainage are desirable. I've heard of several instances over the years of hives being lost to flooding, so this is also an important factor to consider. Also avoid frost pockets.

As for availability of water, I never have to give it a thought, but in many parts of the world it could be critical to colony survival. Bees need access to water for diluting food and cooling the colony in hot weather. If you have to provide water, be sure not to let it run out. A drinker can be any container filled with pebbles for the bees to land on so that there is no risk of them drowning. As the water level drops, they crawl down amongst the pebbles. To minimise bee faecal contamination, site the drinker in a shady spot well away from the hives and not in the bee flight path.

Aim to limit each apiary to about three hives. There are two reasons for this: it avoids stressing the bees by over competition for forage, and high livestocking densities are associated with higher pest and disease incidence. Bees are no different in this respect. If there is another apiary of significant size close by, then consider reducing the number of your hives to less than three, or find a site elsewhere.

At all my hive positions, with the help of a spirit-level I set a recycled paving slab in the ground so that it is not only horizontal but also its upper surface is level with its surroundings. This gives a very stable base on which to place the hive stand. As I generally face the hives in the quadrant between east and south in order to rouse the bees with the morning sun, the orientation needs to be considered when setting the paving slab. Finally, check the levelling of the stand. A level hive is important for straight comb growth relative to the walls and avoiding leakage from contact feeders. A few shims of slate or glass correct minor lists.

You will also have to consider protecting your bees. The apiary should be fenced against grazing livestock. Sheep could rub against a hive and disturb wintering bees, even displacing boxes. Woodpeckers, rats, badgers, skunks, bears, etc. may regard the hive as a ready meal. In most cases the solution is a wire mesh cage round the hive. For example, if woodpeckers are an issue in your area, you will need to net round the hive so they cannot reach the box walls with their bills. Skunk/badger damage calls for hive guards of something more robust, such as a thicker gauge weld mesh. But protecting against bears could make your beekeeping too costly to sustain. And, who knows how many bees we lose to swallows, swifts and martins in summer?

Black knapweed (*Centaurea nigra*)

4 Personal protection and tools

4.1 Protection

Many beekeepers start with all over protection – veil, suit, gauntlets and boots – and if I am opening hives I generally still wear such protection. However, many more experienced beekeepers, especially those practising natural beekeeping, manage with no protection whatsoever. They claim that the 'toxic waste overalls' put too great a barrier between the beekeeper and the bee, resulting in a loss of sensitivity to her mood and intentions. Gauntlets indeed reduce sensitivity when handling combs of bees, but in my experience this is not a problem so long as you are keenly aware of the risk of squashing bees. The gloves most commonly used by UK beekeepers are latex dishwashing gloves. These combine ease of washing with minimal loss of sensitivity, the thinner-walled, tight-fitting nitrile gloves being even better in this respect. After a sting through a latex glove, it can be quickly lifted from the skin, thereby pulling out the sting and reducing the dose of venom. Less sensitive are gloves of plastic-coated cotton or leather.

Beekeeping books advise wearing light-coloured clothing in the apiary, dark clothes, especially of wool, being said to provoke bees to defensiveness. However, I routinely wear navy blue cords in my apiary and the bees seem unconcerned by them. Another common piece of advice it to avoid perfumed cosmetics/toiletries.

I wear a jacket with a built-in veil for brief jobs at the apiary and a full suit and veil for more invasive work. Many beekeepers are very satisfied with a jacket with a good elasticised waist worn over trousers. In hot weather, wearing wellingtons for any length of time is most uncomfortable. Leather safety boots are more breathable. The elasticised bottoms of bee suits fit neatly over them. As mentioned in § 1.5, because of the risk of permanent eye damage, I advise wearing a minimum of a veil when handling bees. Very cheap veil-hat combinations are readily available. Alternatively, black veil material can normally be obtained from fabric shops. This can be worn over a broad-brimmed hat and stuffed in a pocket when not needed. There are veils and veils. I have noticed that the rectangular or square thread pattern on veils of jackets with the concertina/arch pattern of veil support are less transparent than the thinner, hexagonal thread pattern of the more traditional hat/hoop design. If you are looking for eggs in cells, this could be important.

4.2 Small tools

A medium capacity stainless steel **smoker**, preferably with a heat guard, and a **hive tool** are the primary tools of the beekeeper. A chisel at one end and a curved scraper at the other I generally find more useful than the J-type tool. The slim version of the J-type is useful for prising up top-bars that have been pinned, after first freeing the comb from the walls (see below). Much space has been devoted in bee books to **smokers**, lighting and keeping them going. I fuel mine with rotten/punk wood thoroughly dried long in advance. It must be easily breakable with the fingers. Use chunks up to 50 mm (2"). I light a handful of wood shavings with a gas lighter, and drop them into the smoker. Then I throw in chunks of fuel, puffing the bellows as I do so. When a good flame is leaping up, I close the lid and press the bellows from time to time. Aim for smoke that is cool, white, and thick rather than hot, blue, thin, and sparky. (Safety hint: light the smoker *before* putting on the veil.)

Dry newspaper also kindles well. Other fuels: dry leaves, pine needles. Avoid fuels compounded with additives. Observe fire precautions in the vicinity of your apiary. The smoker may be extinguished by stuffing grass into the chimney. Give it an occasional decoke by scraping with a knife, preferably while hot. The first thing to fail is the bellows material. I have successfully repaired mine with car tyre inner tube.

It makes sense to take your smoker, fuel and lighter to the apiary just in case. I rarely even light the smoker for nadiring a new box or inspecting underneath. For more invasive operations including hiving it is advisable to have a lit smoker to hand even if you do not intend to use it. Some advise always smoking the bees in advance of any operation as it is supposed to disguise the alarm pheromone and thus forestall any general reaction. However, I do not find this to be necessary. There is much to be said for a calm, planned, purposeful, gentle, slow manner while keeping well aware of the mood of the bees and biding your time if at first they say 'no!'

Beekeepers' **brushes**, especially those with synthetic bristles, seem to annoy bees. A wing feather or even the outer wing of a large bird such as a goose is sufficient and gentler. (Where I later use the terms 'brush' or 'brushing' as verbs, I do so because the verb 'to feather' has not yet passed into everyday apicultural parlance!)

Three tools peculiar to Warré beekeeping are the comb knife, the cheese-wire, and the comb holder. The Warré **comb knife**, first described by Bill Wood (Oregon), with a blade at right angles to the shaft is not, to my knowledge, purchaseable. Nor is it mentioned in Warré's book. One wonders if Warré ever took individual combs out of a colony in the People's Hive. The first Warré comb knives were made by welding together a stainless steel blade and shaft. Since then, some have been laser cut from stainless steel sheeting. The knife could also be made from a kebab/barbecue skewer, although heating to red heat may be necessary.

However, as the tool need not be made of stainless steel, it could be shaped from 4 mm ($^1/_8$") diameter fencing (straining) wire without heating. Cut a piece 450 mm (18") long and sharply bend it 25 mm (1") from the end to form an 'L' shape, whose short base is to become the blade. With the 'L' wholly in a horizontal plane and resting on a heavy piece of metal, hammer the blade section flat to a thickness of about 2 mm ($^1/_{16}$") and a maximum width of about 5 mm (¼"). It will now be wider at its outer end. File or grind the blade into a roughly rectangular shape and sharpen the *upper* edge. About 100 mm (4") from the other end of the wire, bend it round into a gentle curve about 30 mm (1¼") wide to form a handle. This bend is in the same plane as the blade so that the user can easily determine the blade orientation when it is concealed amongst the combs and bees (Fig. 4.2).

The **comb holder** is a simple stand on which to rest combs fixed to top-bars while

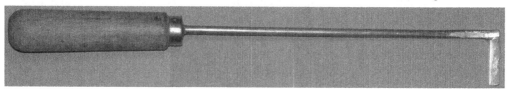

Fig. 4.1 Stainless steel comb knife

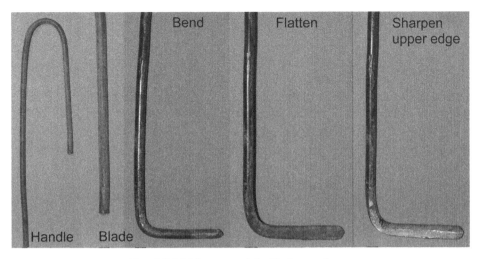

Fig. 4.2 Making a comb knife from wire

viewing the cells. Some top-bar beekeepers use an artist's easel or some similar structure to support combs during examination (Fig. 4.3). An easel can be set on a slope so that the traditional procedure of having the sun shining from behind over one's shoulder can be arranged. This aids quick examination.

A **cheese-wire** and two wedges may very occasionally be useful to have to hand. If the combs in a box become bound to the top-bars of the box below, the wire is drawn through to cut the adhesions at the bottoms of the combs. Although cheese-wires can be readily bought, the tool can easily be made from two short pieces of thick dowel, e.g. from an old broom handle, and a length of steel wire of suitable tensility. I use a single strand from old brake or gear cable off a bicycle. Very thin wire from a piano or other stringed instrument would suffice, as would nylon monofilament or the wire used for hive frames. The thinner the wire, the sharper the cutting. Have spare wire to hand in case it breaks. The **wedges** may be of wood 25 x 100 mm (1-4") tapering from about 13 mm (½") thick to nothing. Two hive tools may substitute for wedges.

Other small tools that may be useful at times include a mirror, small torch/light, camera and weigher, which may be a spring or electronic. An inexpensive digital luggage weigher makes hefting the hive easy. It comes with a hook or a strap or both. The hook is

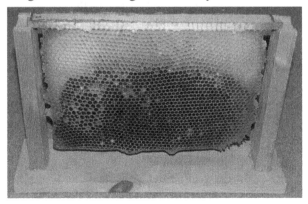

placed under a Warré box handle and, by lifting only about 3 mm (¹/₈"), a reading taken for one or more boxes, the weighing being repeated on the opposite side. If you plan to use a weigher, record in advance the average weight of your hive boxes complete with 8 top-bars, as you will need this figure later.

Fig. 4.3 Comb holder

4.3 Tool kit checklists

Basic	**Extended**	
Smoker	Mirror	Drawing pins
Fuel and kindling	Torch/light	Rubber bands
Gas lighter/matches	Weighing device	String
Hive tool(s)	Water/syrup sprayer	Duct tape
Feather or brush	Perforated newspaper	Secateurs
Comb knife	Queen clip/catcher	Skep or swarm box
Comb holder	Top-bar spacing templates	Et cetera
Cheese-wire	Spare gloves	
Wedges	Sting first aid	
Notebook	Sharp knife/scissors	
Box stand (Fig. 7.2)	Entrance restrictor mesh	

4.4 Lifts

Warré appears to have always had the luxury of an assistant when working his hives. Indeed, the many manipulations performed by himself and his assistant are charmingly

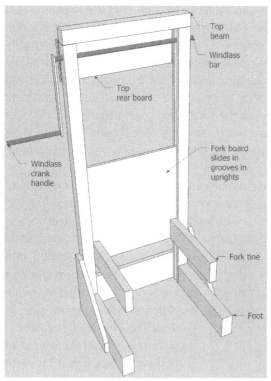

depicted in the photos in an early edition of his book.[10] But if you are without an assistant and have several hives to manage, you may wish eventually to obtain a lift. However, if such a task is beyond your resources you may be reassured to know that all manipulations of hive boxes, including nadiring, can be done by one person (§ 7.2, 8.1, 8.2). Lifts specially designed for Warré hives are not available commercially. But as they can be made by enthusiastic constructors from readily available materials. I describe two here.

As each box has sturdy, projecting handles, it was appropriate to design a fork lift that picks up the hive by its handles. The lift described here, the first of which is believed to have been produced by Marc Gatineau, works on the principle of the French guillotine. A board slides vertically in two grooves and is lifted with a windlass. The groove and the board edges are thoroughly rubbed

Fig. 4.4 Warré hive lift after Gatineau's design (pulleys and cord not shown)

[10] Warré, E. (1923) *L'Apiculture Pour Tous*, 5th edition, Tours. http://warre.biobees.com/warre_5th_edition.pdf

with beeswax before assembly in order to promote a smooth action. Two fork tines are mounted on the board, and the hive box fits between them. The hive can be picked up by the bottom box or, if the lift is tall enough, by any of the boxes above it. A mini, and therefore lighter, version of the Gatineau lift, designed to sit on box handles, is also described.

4.4.1 Gatineau lift for the Warré hive

No imperial measurements are given in the plans for this lift, as much will depend on the thicknesses of the wood available in a particular locality and on the width chosen for the hive boxes. Suffice it to say that the uprights, feet, forks and top beam are about two inches in thickness and the plywood boards about half an inch. In the plans, the clearance between the fork tines and the walls of the box is 8 mm ($^5/_{16}$") for a box width of 340 mm ($13^5/_{16}$").

The fixing of the fork tines to the rear board requires careful attention. On the bottom of the back of the 13 mm (½") thick fork board is glued centrally and flush with the bottom edge another plywood board measuring 412 mm x 200 mm x 13 mm (16¼ x 8 x ½"). This is to reinforce the fork board in the region of the tines.

The tines of the fork are cut perfectly square at the ends, butt-jointed with glue onto the board and clamped in place at right angles with the help of supporting blocks. When the glue is dry, two 100 mm (4") wood screws are screwed into the

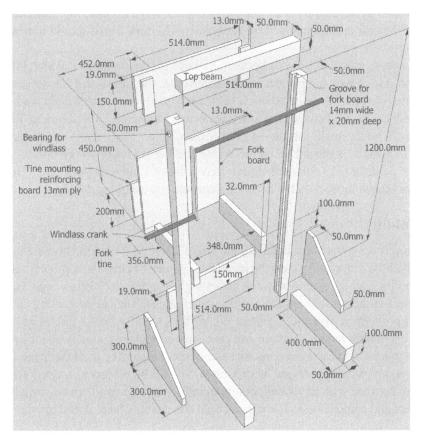

Fig. 4.5 Exploded view of a Warré hive lift

tines through holes drilled and countersunk into the back. The screws are size 12, i.e. 6 mm
(¼") thick at the widest end. The screws are lubricated with wood glue when inserting so
that any play is taken up by the glue as it sets. The extra board glued on the back takes the
vertically sliding unit ('guillotine blade') to a thickness of 26 mm (1") and confers greater
rigidity. It must of course clear the inner edges of the uprights. The clearance shown here
is 1 mm each side.

Other successful methods of fixing the tines include sturdy shelf brackets (Andy Collins,
UK; Steve Ham, Spain) and 'T'-hinges (Bill Wood, USA).

The top pulley is tied to the top beam and the bottom pulley to the fork board. The
lifting cord is tied to the windlass axle, threaded through the two double pulleys and tied
to the fork board. Pulleys are not essential, but they do ease lifting of a tall, honey-laden
hive and allow movement to be stopped at any point without securing the crank. If your
lift is very free-running, for safety's sake include a windlass lock. The top rear board is
set out on 13 mm (½") spacers to clear the windlass cord and top pulley.

The windlass and crank handle could be made of steel pipe used for example as
electrical conduit , or reclaimed from other machinery such as cars, mangles, scrapped
honey extractors, etc. To increase durability of the windlass bearings, they may be lined
with steel bushes and greased.

No effort has been made to reduce the weight of the lift, which is somewhat heavy
(12 kg; 26 lb), but several dimensions could be reduced by a few millimetres without
significant loss of rigidity, for example the thickness of the upright, top beam and feet as
well as of the rear and triangular bracing boards. The fork board should not be thinner
than 13 mm (½").

Useful accessories that various beekeepers have incorporated in a Warré lift include
wheels and a hive weighing scale. In the latter case, the fork board has to rest against
roller bearings in order to reduce friction sufficiently to give accurate weight readings. The
ability to adjust the gap between the fork tines would be a helpful development for picking
up hives with boxes made with wood thicker than 20 mm (¾"). An extraordinarily wide
range of lift designs is illustrated on the Warré beekeeping English web portal.[11]

I briefly describe one other lift design here, that of Raimund Henneken (Germany),
because it is much more portable and can incorporate a weighing device.

4.4.2 Mini-lift

This version is designed to lift either all the boxes, by resting it on the hive stand, or only
some of the boxes by resting it on the handles of any box below the top box in the stack.
One reason for opening seams between boxes higher up the stack is to weigh the top one
or two boxes in the stack when estimating winter stores. The use of bearings represents
a major improvement over the Gatineau lift, with its wooden board sliding in a wooden
groove. This greatly reduces friction and gives two advantages. One is that it avoids any
risk of the juddering during lowering that some report regarding their Gatineau lifts. The
other is that any weighing device incorporated will not be subject to serious errors. The
lifting may be done with any small, commercially available hand winch, which is ideal
for lightness and compactness. The winch could conceivably be replaced by a home made

---

[11] http://warre.biobees.com/lift.htm

Fig. 4.6 Steel mini-lift. Photos: Raimund Henneken

Fig. 4.7 Mini-lift with drawer runners, shelf brackets and weigher. Photo: Andy Collins

windlass with pulleys, but this would add to the height and possibly to the weight. This lift weighs 4.2 kg (9¼lb) and was tested with a 60 kg (132lb) load.

Andy Collins (UK), has designed an equally compact lift which, apart from using adjustable shelving brackets/ supports for the feet and forks, also minimises friction by incorporating kitchen drawer runners. This version does not have the fork tines directly above the feet, and is therefore not designed to rest on the handles of the boxes.

5 Getting bees

Warré advised populating the People's Hive with a swarm of 2 kg. (4½lb) or more (p. 70, *Beekeeping for All*). However, many Warré hives in the USA have been successfully populated with 2-pound (1kg) commercial packages. Package bees comprise an artificial swarm with a queen and some candy, all in a box fitted with mesh ventilation. In this section, I cover obtaining bees either on or off comb.

Starting Warrés with bees off comb rather than on comb is usually the easier of the two options. Furthermore, a swarm not only leaves behind the greater part of the pests (e.g. varroa) and pathogens of the parent colony, but also, if it is a natural swarm, comes ready-fuelled and very eager to set to work building its nest of combs. On the other hand, bees on comb can be minutely inspected for symptoms of disease before sale/purchase. On page 67 of *The Bee-friendly Beekeeper*, I compare vertical pathogen transmission (via a swarm) with its horizontal (via brood comb) counterpart. Repoduction via swarming is more hygienic, and selects against pathogen virulence.

There is a lot to be said for using bees that are adapted to your local climate, forage, and pest/pathogen spectrum. Furthermore, imported colonies with pedigree queens have sometimes given rise to increased defensiveness after second generation crosses with local bees. However, some people will have no option but to buy commercial packages or bees already on combs.

If you start your Warré with a package, or bees on comb (a nucleus or 'nuc') you will certainly have to buy them, in which case order in the winter or as early as possible in the season or you may be disappointed. Package bees may have travelled tens of thousands of miles, but, surprisingly, that does not necessarily stop them making prosperous colonies. However, try to get them from as near you as possible.

Some frame hive beekeepers, on discovering the Warré hive, immediately want to transfer their bees from their frame hive to a Warré hive. In this case it is worth addressing a few questions: is the transfer really necessary? Put another way, is it worth disrupting a perfectly healthy colony in a frame hive just for the sake of trying out a Warré hive? Would it be possible to wait and see if the frame hive issues a swarm? I have populated two Warré hives from frame hives, but only because the frame hives were on the point of swarming, i.e. had many occupied queen cells, and, at the time, I was still managing swarming in my frame hives by the artificial swarm method. The procedure that I used is described in § 6.3.2.1.

5.1 Swarms

If you are determined to try for a local swarm, and do not mind the risk of not having bees at all in the season when you get your hive, then here are some suggestions for obtaining swarms at swarming time, which is mainly April to July in the northern hemisphere:

1. Join your local beekeeping association and make it known that you are looking for a swarm or artificial swarm. Advantages: you could get locally adapted bees and pay much less than full commercial rates, possibly nothing at all; you may see practical demonstrations of taking swarms. In UK law, a swarm you take is yours, but for access you must first ask permission of the landowner.
2. Tell your local services that receive swarm calls, e.g. local Council pest control, police, etc., that you will take swarms.

3. Set up bait hives, sometimes called 'swarm lures' (see below).
4. Post a message to your local internet recycling group (e.g. Freecycle), or put a card in the window of the newsagents.
5. Create a web page advertising that you will collect swarms within a radius of travel that you specify.
6. Tell local pest control companies that you will collect swarms. Many do not like to have to destroy honey bees. If your travelling costs are an issue, bear in mind that clients will sometimes prefer having bees removed alive and will pay a fee.

Happily, bees are generally at their most docile when swarming. The only sting I had from the dozens of swarms I have taken was when a swarm fell on my head and shoulders while I was perched up a ladder. Even so, when taking swarms in public places, or on other peoples land, consider what safety precautions are required and, in these times of increasing litigiousness, whether you need public liability insurance, or even written indemnification from the landowner regarding consequential damage, especially if some cutting will be required.

If you can take most of the bees and the queen, you are almost home and dry.[12] I do not try to list here all the possible ways of taking swarms, but instead I describe a few of the more common ones. The minimum you need is a cardboard box with about a 300 mm (1 ft) square opening and a cotton cloth to cover it. I still use the same box with which I started taking swarms in 2004. The traditional container for catching swarms is a skep, but one really need not go to that expense. Other useful

Fig. 5.1 Two swarms for taking; one easy one more difficult

---

[12] Natural and biodynamic beekeeper, Johannes Wirz, accurately describes the queen as the 'organ of colony coherence', preferring that to her usual designation as an egg-laying machine.

things to have if you have time to fetch them are a veil, smoker, piece of brood comb, and secateurs or lopping shears. A ladder may be needed, but please do not join the long line of beekeepers injured by falls from ladders!

If the swarm is in a bush or on a tree branch, clip away twigs and foliage to allow the opening of the box or skep to be placed directly below it. Have a cloth draped over the arm holding the box. With a sharp blow or shake with the other hand, knock the swarm into the box, being prepared for the extra weight which could be over 3 kg (6½lb). Mayhem may seem to break out, but ignore it and immediately cover the opening with the cloth then gently carry the box to a shady spot close to the swarm site. Invert it slowly onto the cloth and place a stone or stick under the rim to create a narrow entrance. If the queen is in the box the bees that have taken flight will scent her and enter the box. If all the bees leave the box you will have to repeat the process once the swarm re-clusters. Ideally, leave the swarm in the box until evening when all the bees have returned. This would be especially important if the swarm contains a virgin queen who might go on a mating flight before evening. You may then move it and hive it. If no shade is handy, improvise with a sheet, board, leafy branches, etc. The swarm may be moved up to a few miles in its container loosely wrapped in a sheet or sacking with space underneath to allow ventilation.

Warré suggested knocking swarms directly into a Warré box. This is very hard to do single-handedly, as my boxes have 25 mm (1") thick walls and are too heavy to hold in place with one hand, especially while up a ladder!

Frequently, swarms are not in places where a knock- or shake-down is possible. For example, for a swarm up on an ivy-clad tree trunk (Fig. 5.1b) I did as follows: taking the brood comb, which on that occasion was supported in a frame, I held it against the top of the swarm. Bees rushed onto it. When it was really heavy with them, I carried the cluster, without jarring it, to the box on the ground and shook the bees in, inverting the box slowly on a cloth and resting its edge on a stick. I repeated this a few times. Eventually all the bees in the tree took to the air and entered the box.

Another method is used for a swarm on a fairly even surface, such as a fence, post, wall or tree trunk. The box, which could also be a ready prepared Warré box, with a top-bar cloth and piece of cardboard stapled over it, is supported with its bottom opening immediately above the swarm and touching it. Bees perceiving a dark cavity begin to ascend into it. They can be encouraged upwards by gentle brushing with a feather. Rarely does one need to start them off with smoke, and then only with few light puffs. Drumming the support on which they are clustered, followed by a pause, then more drumming, may also encourage them to run upwards.

I hive swarms in the evening of the day they are taken and only very rarely earlier or the following day (§ 6.1). Ideally, hive early in the season, say at the dandelion flow, as this gives plenty of time for a colony from a swarm to establish itself before winter. The old saying amongst beekeepers has much truth in it: a swarm in May is worth a load of hay; a swarm in June is worth a silver spoon; as swarm in July isn't worth a fly. However, a late swarm, given a box of comb and plenty of honey, could still build up to wintering strength in time.

If you come upon a swarm late in the swarming process, it may already have made the decision as to where its new home is to be. The 'buzz runners' may be preparing it for flight. That is why speed is called for once you find a swarm. Therefore, at swarm time keep your swarm-taking kit handy. If someone elsewhere calls you about a swarm, enquire about its precise location, access/permission, height above ground, length of time it has been there (a day-old swarm 20 miles (30 km) away is not so alluring) and, most importantly, whether the person who saw it was not looking at a colony of wasps. It does happen! Once, I asked for a photo of the bees seen by a potential client to be sent me by email. It saved us both the cost of 60-mile round trip, as I was able to identify the insects as miner bees.

As this book is primarily intended for people relatively new to beekeeping, I will not discuss here the complicated and lengthy process of extracting swarms of honey bees that have established themselves in places where they are not wanted, usually in buildings. However, as the reader gains beekeeping confidence, he or she may wish to consider this option for getting feral/wild bees that are locally adapted. Clients will pay for the work to be done. Public liability insurance may be advisable.

5.2 The nucleus (nuc)

Obtaining bees already on a few combs, i.e. as a nucleus (nuc) presents the Warré beekeeper with a more difficult problem. In the case of a Warré nuc it would be necessary first to find a bee supplier who would be willing to establish one in a Warré box and possibly even supply the box and other hive parts. Then comes the problem of transporting a box of new comb, which tends to be somewhat fragile. In the case of nucs on frames, transport is relatively easy but transferring the nuc to a Warré hive is problematic.

5.2.1 Warré nucs

At the time of writing, I know of only a couple of instances of beekeepers supplying Warré nucs. You may have to provide the producer with a Warré box with prepared top-bars. The producer will shake bees into it, give it a queen and very likely help its development along by feeding sugar, or, if you are lucky, honey. At some point you have to collect it and transport it to your apiary. It is unlikely that you will want to take your nuc apart and inspect it before taking it away. However, if the seller has already rendered the combs relatively inspectable by freeing side connections, and you wish to inspect, some key quality criteria are listed below (§ 5.3). New combs are very prone to detach from top-bars even when observing ideal transport criteria, which comprise: a mesh screen top and bottom to maximise ventilation; the bottom screen well clear of any surface on which the box rests; the combs parallel to the direction of travel, and the actual move done in the cool of the evening, or at night. If it is nevertheless warm, it helps to *lightly* spray the top mesh with water now and then.

To minimise the risk of comb detachment in transit, a Warré nuc could be started with four or five frames instead of top-bars and the nuc shipped only when the frames are filled with comb. The frames of comb need not be started with foundation; a starter strip of wax would suffice. Ickowicz (France) supplies Warré boxes with frames that have a slot in the top-bar for inserting/securing foundation. Simple frames are easy to make provided one has access to wood of about 7-9 mm (¼-³/₈") thickness and a table saw to cut it to width. They need not be so precise that they have a perfect bee space round them, for, if the bees fix them in place with propolis or comb, they will anyway be removed at the first full harvest. If you buy Warré boxes, check that they meet the specification of a standard Warré hive, or, if they do not, whether any discrepancies might inconvenience you later. However, you could always stick to equipment from one supplier.

5.2.2 Nucs on frames

Nucs on frames (National, Langstroth, etc.), probably the commonest way bees are sold in the UK, whether by a local beekeeper or by a breeder further away, are probably the least practical start for a Warré colony, whichever way one chooses to make the transfer. Somehow the colony has to be persuaded to occupy the Warré (§ 6.3.2). Not infrequently the bees are slow to do so without radical 'surgery'. Beekeepers resort to this option if the options of neither a swarm nor a Warré nuc are available to them. Nucs, normally already supplied in well ventilated containers, are transported with combs parallel to the direction of travel. In hot weather, an occasional light spray of the top mesh with water is advisable.

When buying a nuc be sure it has a good quality, young laying queen who is the mother of all the brood. You may wish to specify in advance that you do not want her mutilated by marking or wing-clipping. Three frames with disease-free brood at all stages, including eggs, should be present, and at least four frames fully covered with bees. In aggregate, there should be a comb of honey and half a comb of pollen stores. The combs should be light coloured showing that they are less than a season old. Your state's bee health department may give additional advice.

5.3 Bait hives

A bait hive (swarm lure) can be any container that smells of bees, has the right size and is positioned to optimise attracting a swarm that is looking for a home. Seeley and Morse[13] studied this process and found that the bees had the following preferences:

- nest height at 5 metres preferred to1 metre (16 ft preferred to 3 ft)
- entrance area of 12.5 cm^2 preferred to 75 cm^2 (2 sq.in. preferred to 6 sq.in.)
- entrance position: bottom preferred to top of nest cavity
- entrance orientation: southward preferred to northward
- nest cavity volume: 40 preferred to 10 or 100 litres (10 US gall. preferred to 2½ or 25 US gall.)
- cavities previously occupied by bees preferred to new cavities
- cavities more than 300 m (325yds) from the parent colony preferred to ones very near it
- more visible sites preferred to less visible sites

Although bait hives placed to attract your own swarms should ideally be some distance from your hives, it is worth noting that bees do occupy bait hives in the same apiary as the colony from which they issued.

To make the cavity smell as if it has been occupied by bees one can use a hive or part of a hive, for example, a propolised Warré top-bar cloth. Some rub the inside with beeswax. More effective is a piece of empty honey comb or brood comb which must be clear of

[13] Seeley, T. D. & Morse, R. A. (1977) Dispersal behaviour of honey bee swarms. *Psyche* **84**(3-4), 199-209.

clinical signs of brood disease. If in doubt, do not use it. Instead of scenting the cavity with something from bees, some beekeepers use a *couple* of drops of *Cymbopogon citratus* oil, or a synthetic swarm lure pheromone which is obtainable from apicultural suppliers in plastic vials which are pinned inside a bait hive.

The cavity may be anything from a cardboard box to a 2-box Warré ready set up to house a colony. If cardboard is used it must be protected from the weather and the swarm may not be left in it for long as the bees begin to chew the cardboard. The most entertaining bait hives I have seen are cabin trunks situated on balconies of high-rise apartment blocks (see Fig. 13.1, p. 85). A cheap and light bait hive that can easily be hung on trees etc. is made from two natural fibre plant pots tied together. The combined volume may be smaller than preferred by bees but, as Seeley and Morse found, even 10 litre (2½ US gallon) cavities were not entirely rejected if they met other criteria. However, some fibre pots have capacities up to 15 US gallons.

In spring, I scatter about 15 bait hives around the district, often sited on walls and roofs. Several entrances are visible to landowners as they pass and they alert me if scout bees are seen. Otherwise, I check on a daily tour.

From the swarm's point of view it is better if it can build its nest uninterrupted from the time it arrives in its chosen cavity, and remain at the same site. This would apply to a bait hive at its final location in an apiary. It is even said that colonies started this way do better than those re-hived elsewhere, or even just relocated. However, I have not tested this. In most cases it will be impracticable to leave the occupied bait hive, so re-hiving or relocation will be necessary.

Once a colony is in a bait hive, it is essential to move it to its final location on the day it arrived. Otherwise, the bees orientate to the bait hive and moving them to a site inside a radius of three miles could cause the colony to lose many bees. On one occasion a swarm that was taken from me by another beekeeper absconded and returned over a distance of a mile to my bait hive. This was confirmed by the bee-

Fig. 5.2 Warré-hive-sized bait hive and swarm transporting box

keeper seeing the direction in which it flew off, and my finding the hive reoccupied a short while later by a swarm of the same weight as the one he had received. Another reason for dealing with bait hives the same day is that a swarm rapidly builds and fills comb which would be wasted if the bait hive is only a temporary affair.

If you have been a little late in relocating your incoming swarm, the following method of recovering 'lost' bees has worked for me. When the swarm is hived, at dusk place grass very loosely on the alighting board and put a few branches round the entrance. These obstacles cause bees emerging the following morning to re-orientate to the new entrance configuration and hive site. But some will still go back to where the bait hive was. So at that site place an empty hive or other suitable shelter, which could be the bait hive itself. At nightfall, move it to the rest of the swarm now hived, and place the two entrances as close as possible to each other. Lost bees emerging from the temporary shelter scent the parent colony and enter it. This may have to be repeated once or twice.

A versatile design of bait hive for Warré beekeepers is a light plywood box, the shape and internal dimensions of two Warré boxes joined together, fitted with two handles and containing a removable grid of eight waxed top-bars that are ready spaced to be lifted as a unit and lowered into a Warré hive. The lightness allows the hive easily to be hung in a tree or on a telegraph pole. Additional features include an entrance closable with a mesh, and openable, meshed ventilation holes top and bottom. Such a bait hive box also serves for taking and transporting a swarm. As it is not always possible to tell at a glance from entrance activity that a swarm has gone in, a useful additional feature is a small window near the top of the back. The design of the box allows a swarm to be moved a considerable distance, even during the day and, once at the chosen site, it could be left in the box for a few days before final hiving.

In some countries hives have to be registered with the bee health authorities against payment of a fee and even bait hives are included in this. The resulting bureaucracy and cost may make this method of obtaining bees less attractive.

Fig. 5.3 Swarm just arriving at a bait hive above the author's porch

6   Hiving

I like to hive bees in the hour before sunset. Conditions are cooler and duller. The bees are less likely to abscond at this time. I have only once had a problem with late evening hiving and that was when, after some delay, dusk came on quicker than desired, and many disorientated and angry bees clung to our clothing. So be warned: do not leave it until dusk.

Here is a basic checklist for before you set off to your apiary: hive site/stand prepared; complete hive with 2 boxes; top bars waxed and fitted in both boxes; top-bar cloth sized; basic tool kit plus sugar solution (1:2 sugar:water by volume) spray/mister and top-bar spacing template (hiving packages); veil; sheet & board (run in only, board dimensions roughly 60 x 90 cm – 2 x 3 ft); pliers; funnel or Warré box without bars (two boxes in the case of packages); bees in swarm container or package or nuc (Warré or framed). See § 6.3.2 for additional equipment for hiving nucs.

With swarms, the two options are shake/pour in or run in. I find the run in the most enjoyable, but it is usually the slower method of the two as it can average about half an hour.

Boxes may be orientated 'cold way', i.e. with the combs/top-bars at right angles to the side with the entrance, or 'warm way' with the combs parallel to the entrance. I will leave the reader to work out the full meaning of these old beekeeping terms. Although Warré advised using cold way in summer and warm way in winter, I use cold way all the year round on the grounds that the bees have no general preference for comb orientation with respect to the entrance, and certainly do not rotate their combs through 90 degrees in autumn and spring.

An ideal time to hive in the northern hemisphere is just as the dandelion flow is getting under way. However, if it must be done in less than ideal foraging conditions or weather, consider whether you need to feed. If feeding is necessary, for bees not hived on comb, feeding at the top of the hive is advised (§ 11.1).

If your bees are in a district prone to foulbrood, and you intend to hive swarms of unknown origin, some biosecurity precautions would be prudent. Bees leaving a colony where symptoms of foulbrood are manifest may have large numbers of spores of the pathogens in their crops together with the honey they are carrying for the journey and for setting up their new home. A precaution would be to allow the bees to draw combs for about three days then re-hive them with a box with fresh top-bars and burn the combs already built. A further measure would be to quarantine the colony, preferably in a quarantine apiary, and when it is well developed inspect its combs for signs of foulbrood.

6.1 Swarms

If you took your swarm in a Warré box or two, then there is little more to be done but to set it on the floor and stand, with a minimum of two boxes, and fit the quilt and roof. Also see § 6.1.3.

6.1.1 Running in a swarm

Set up your hive in its final orientation with two boxes and their waxed top-bars. Put in place the top-bar cloth, quilt and roof. Place the board to slope up from the ground to the alighting board. Cover it with the sheet and arrange the sheet on the ground both sides of the board to catch any bees that miss the board. Weight the edges of the sheet with pebbles if it is windy.

Remove the covers from your swarm and with a sharp but not too violent shake, dump the bees near the top of the board right in front of the entrance. They will spread out like treacle and some will take to the air. Shake or brush out of the container any remaining bees, or place it to one side opening towards the swarm. You may wish to mist the wings of those on the board with syrup water if excessive taking flight occurs. However, usually this is quite unnecessary, as the flying bees soon rejoin the swarm. Within a few seconds of being dumped, a kind of 'roar' goes up from the crowd and bees are seen rushing in at the entrance. Many Nasonov glands are exposed and the lemony scent of the 'it's here girls!' pheromone may be detected. If you watch carefully, you may be lucky enough to see the queen running up over the backs of the workers.

Fig. 6.1 Swarm running into a hive

The quickest hiving I have done this way lasted about 10 minutes. The slowest has required me finally to go home at nightfall, leaving a big cluster of bees hanging from the alighting board. But by morning they were always all inside. Rhythmic tapping the board for a short while followed by a pause can speed up the run-in. You may also use a feather to scoop bees towards the entrance. Smoke can get little groups of stragglers moving, but it can also disorientate them. Eventually you can remove the sheet, and, a little later, the board.

It rarely happens that the cluster does not hang in the top box and start building there. If it does start in the box below, merely change the places of the two boxes, keeping the orientation with respect to the entrance unchanged.

An artificial swarm in a package may be hived similarly either by direct release of the queen at the entrance *immediately* prior to dumping the bees on her, or by hanging her cage from the top-bars of the top box after removing the cork covering the candy plug. Opening and handling packages is detailed in § 6.2.

6.1.2 Shaking/pouring in a swarm

Have a lit smoker to hand just in case. Set up your hive in its final orientation with two boxes and the waxed top-bars arranged 'cold way', i.e. at right angles to the side with the entrance. Have the top-bar cloth, quilt, and roof ready at hand. On top of the hive, put a Warré box without top-bars, or a funnel of your own design, which may even be of cardboard covered with plastic to make it slippery. Remove any covering from your swarm container (box, skep, etc.), invert it, give it a sharp thump or two to disperse the cluster and pour the bees into the hive like coffee beans. Thump the container again to detach more bees and pour them in. Place the container before the entrance so that the few bees left will find their sisters and enter the hive.

The bees will gradually sink below the top-bars. Brush down any still on the top box or funnel. Light application of the water spray/mister or smoker at this point may occasionally be necessary. When all but a few bees are down, replace the box/funnel with the top-bar cloth. The bees, not liking it resting on them, crawl down between the bars. When the cloth is on the bars and there are no bees trapped, put on the quilt and roof.

6.1.3 Absconding

When, because of other commitments, I have had to hive swarms in the afternoon, occasionally they have absconded. In one case I had to repeat twice the taking and hiving. Absconding could be triggered by overheating in the hot afternoon sun, or possibly by the bees not liking the smell of a brand new hive.

Apart from hiving towards sunset, a useful precaution to prevent absconding is to place below the cluster a queen excluder, which here acts as an *includer*. This must be done once the bees have entered the hive and the bulk of them is with the queen hanging in the top box. Especially in the case of a run in, if it is done too soon, the queen could be trapped below the excluder. When the colony has settled after hiving a swarm, remove the roof and *gently* lift the hive and quilt off the floor, setting it aside. Place on the floor an eke, which here may be anything from a frame of wood about 25 mm (1") deep to a hive box without top-bars, and put a queen excluder on it. Put the hive and bees back on that and replace the roof. Leave the includer in place no longer than a *couple* of days. If I use one, I usually remove it about 24 hours later. If the swarm is small (cast, after-swarm, secondary swarm), it could contain a virgin queen who needs to be able to get out to mate.

Once a swarm has used up the stores it brought with it, usually between two and three days after issue, the risk of absconding is reduced, as the swarm lacks the fuel to move and establish itself elsewhere and will usually have committed itself to the new abode. Therefore, avoid feeding until this period is over.

6.2 Packages

As the beekeeper has usually invested much money in a package of bees, he or she may wish to make extra certain they know in advance exactly what to do when hiving the package, if necessary rehearsing it with a dummy run beforehand and/or watching videos of the procedure on the internet. The worst, though rare scenario is losing or killing the queen in the process. But even this can be solved by immediately buying a replacement queen, or by getting a spare one from a friendly neighbouring beekeeper. For the following instructions, I thank John Moerschbacher (Canada) who hived thousands of packages in Langstroth hives before trying Warrés. For the Warré he describes two methods involving either direct or indirect release of the queen, and the slight variation required if one wants to have the finished, populated hive with *fixed* top-bars.

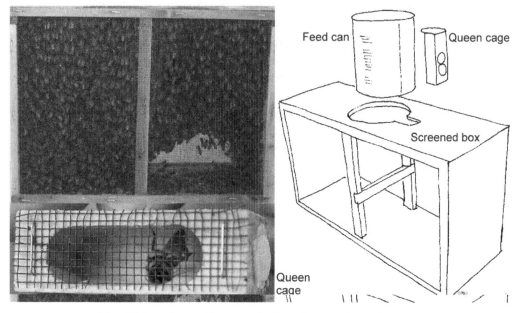

Fig. 6.2 A package of bees, including closeup of caged queen

6.2.1 Direct release of the queen

A commercial package (artificial swarm) can be hived by direct release of the queen, if she has been with the bees for at least two days. As an additional tool for hiving you need a 3" (75 mm) nail.

Spray the bees gently with thin tepid syrup every half hour for a few hours before they are hived. Do not soak them. If you tilt the cage a little and just coat the screen/mesh, that is enough. Allow them to take in the syrup and clean each other off. You will see thousands of little tongues licking the screen clean. Do not feed too much or too often until a few hours before hiving. Feeding too much too soon can result in a lot of wax build-up around the feed can, making it more difficult to remove.

Before hiving the bees, they should be kept in a room with dim light or dark at about normal room temperature or a little below, especially if the outside temperature when you hive them is below 13°C (55°F). This helps raise the bees' core temperature ready for the cold conditions in which they may be being hived. When you hive them, they will be active enough to cluster up high, even if the outside temperature is very low. When you hive a package in very warm weather, keep the package in a cool dark room for a few hours to calm them. No spraying is needed until just before hiving.

Whether hiving in warm or cold weather, light spraying is recommended. This is not to mask the scent of bees or queen, which come from different colonies, but simply to divert their attention from the trauma that is happening to something more to their liking. Even if your hive will end up with fixed top-bars, you proceed initially as for loose top-bars.

6.2.1.1.Top-bars not fixed/nailed

Put two empty Warré boxes (i.e. with no top-bars) on the floor (bottom board). Jolt the package to knock the bees to the bottom of the cage. Use the nail to prise up the rim of the feeder can so as to be able to get a hold on it to remove it. If there is any difficulty, pull up the can with pliers until you can grasp it. After the can is removed, spray the bees lightly with syrup. Remove the queen cage and carefully prise out the cork, immediately putting your thumb over the hole. (There is seldom if ever a candy plug. One usually only finds candy plugs in so-called 3-hole queen cages in which queens are shipped individually with some worker attendants.)

Place the queen cage, screen up, on the floor of the hive. After taking the thumb off the hole, *immediately* dump the bees from the package on top of her. Use a quick rocking motion to get 99% of the bees in. Then knock the package box in such a way as to 'roll' the remaining bees into a ball in one corner. This ball can be easily rolled out of the hole and into the hive. Spray the bees again, then put the top-bars of the top box in place, and add the top-bar cloth, quilt and roof. A top-bar spacing template is useful here.

Closing up the hive could include putting a top feeder in place, especially if weather conditions or lack of available forage call for it (§ 11.1). If hiving in cold weather, the source of feed must be within easy reach of the cluster hanging from the top-bars, or, within a few days, you could have starved bees. Therefore, a floor feeder in these conditions is not acceptable.

After a day or two, gently lift aside the top box with its cloth and quilt onto a box stand. Remove the now empty queen cage, install the top bars in the lower box and replace the top box. The top-bar cloth and quilt need never come off the hive again until harvest time, unless a top feed is given or it is discovered in the ensuing week or so that the queen was not accepted, possibly through too short a time in transit to let the colony grow accustomed to her. Symptoms include a dead queen at the hive entrance, no pollen pouring in after a couple of days, lack of a tight cluster in the top box, and little comb construction.

6.2.1.2 Top-bars fixed/nailed

This procedure differs only slightly from the above. You still dump the bees into two empty boxes (i.e. with no top-bars), but once in, you simply set a third box on top with top-bars fixed in their proper spacing. Then when you go back in a day or so, you replace the bottom two empty boxes with one box with top-bars. The extra distance the bees will have to crawl up to cluster will in no way be a problem if they are warm to begin with.

6.2.2 Letting the bees release the queen (indirect release method)

An alternative method with a package relies on placing it on end inside the bottom hive-body box with its bars removed. Place on the hive floor (bottom board) two small sticks, about a bee space or more thick, then put one box without top-bars on the floor. Firstly, remove the feed can and queen cage from the package (§ 6.2.1.1) and pour most of the bees into the hive box. Place the package with its remaining bees on end on the sticks, taking care not to crush any bees. Place two boxes with top-bars on the bottom box, the extra box ensuring the cluster can hang above the package box without obstruction by it. Remove the plug over the candy/fondant in the exit hole of the queen cage (or, if there is no candy, as is commonly the case, substitute a soft candy or fondant plug for the cage plug, *taking care that the queen does not escape*) and attach the cage to the bars of the top box so that neither the mesh side of the cage nor the candy hole is covered. Close up the hive. On the following day, when the cluster has formed round the queen in the top box, remove the bottom box, package and sticks. If pollen is not coming in after a couple of days, it is as well to check that the bees have released the queen and, if not, release her yourself.

It rarely happens that the cluster does not hang in the top box and start building there. If it does start in the box below, merely change the places of the two boxes, keeping the orientation with respect to the entrance unchanged.

If you need to replace the queen, indirect release is mandatory as, unlike with a package, she will not have had a couple of days in transit for the bees to get used to her, and they will very likely kill her (see § 7.7).

6.3 Nucs

6.3.1 Nucs in Warré boxes

The hard work is already done for you! Set up your hive with one box with its top-bars on the floor and the top-bar cloth, quilt and roof at the ready. Remove the travelling screens from the Warré nuc and place it in the right orientation on the bottom box. Because of the small size of the colony and the simplicity of the manipulation, you are unlikely to need to use smoke, but have your smoker lit, just in case. Consider whether feeding is necessary (§ 11). Close up the hive.

6.3.2 Nucs on frames

The two options for getting the colony into a Warré are 'growing down' or 'chop and crop'. The latter method is so drastic that I regard it as a method of last resort. However, as it transfers the colony to the Warré in a single operation and no further manipulation is necessary, some might consider it the less intrusive of the two methods in the long run.

6.3.2.1 Growing a nuc down

To grow a nuc down, you need to make a transfer box that fits the frames and sits on a specially made adapter board on the Warré hive. A temporary roof is arranged over the whole, and, as the process could take a year, the nuc box may need insulating if made of thin wood. If the nuc box supplied is non-returnable, you may be able to leave the frames/ combs in it and use it as supplied, removing any travelling mesh from the bottom. It helps to take off the bottom bars of the frames so the bees can extend the combs downwards and thus closer to your Warré top-bars.

The idea is that the nuc colony grows down and you eventually harvest the nuc combs, filled with honey, just as if it were a Warré box. But frequently, especially in nectar poor areas, the brood nest has not sufficiently moved down by the end of the season, so the beekeeper is faced with protecting a precarious and thermally inefficient structure

Fig. 6.3 Transfer box and adapter

from the winter weather. To speed up the moving down, you may use a somewhat more disruptive method that combines a shake/brush down with the use of a queen excluder. When there is at least one comb in the Warré, all the bees are shaken/brushed down into it, or, if you are confident in finding the queen, she is caught in a queen clip and added to the rest of the bees at the end of the shake down. You may need to make a simple funnel for the shake down (Fig. 6.4) or at the very least use an empty Warré box as a funnel.

Remove the transfer box and support it at a convenient height immediately beside the Warré so there is no risk of dropping the queen on the ground. Remove the adapter board and set on the Warré a funnel or a Warré box without top-bars. When using a box, it is unlikely to accommodate a nuc frame, even if the frame is held across the diagonal of the box, so there is a risk of bees falling outside the box, worst of all the queen. So there are two alternatives. The first involves holding the frame vertically by the uppermost lug, lowering it somewhat into the Warré box and, while

Fig. 6.4 Funnel for shake/brush down

gripping the lug tightly in one fist, bringing the other fist down sharply on the one gripping the frame lug. If you are not confident that you will not drop the frame in this procedure, it would be safer to brush the bees into the box while holding the frame the same way. When shaking into a funnel with the frame held horizontally, a sharp shake downwards will dislodge most bees. The shaking is repeated once or twice until there are no bees on the comb.

The shake/brush down is done frame by frame and *rapidly* to minimise chilling the brood, the box/funnel removed after brushing down any bees still in it, the adapter replaced on the top of the Warré hive, a queen excluder placed on it and on that the transfer box with all the brood frames replaced in the same order as they were before. The temporary roof is replaced. Attracted by brood pheromone, the nurse bees now move up into the transfer box to re-cover the brood which continues to develop and hatch out. It is unlikely that there will be much drone brood at this stage, therefore providing a drone escape is unnecessary. After three weeks, all the brood will be hatched out. The transfer box, adapter board and queen excluder are now removed and the Warré closed up in the normal way.

The shake down described above may be used for artificial swarming into a Warré hive a frame hive colony that contains occupied queen cells and might otherwise swarm. Combs with queen cells containing larvae should be brushed, not shaken. A 2-box Warré is placed on the site of the frame hive, facing the same way. After the shake/brush down, an adapter to fit the brood box with frames is placed on the Warré followed by a queen excluder, the frame hive's brood box and its covers. After leaving this combination for only a few hours (2 minimum) to repopulate the combs with bees, the frame hive is removed as far away as possible to another site in the same apiary. Any flying bees automatically enter the Warré hive that is now on the site that they have grown accustomed to and join the queen. The frame hive colony finishes raising new queens and carries on its life.

Warré describes at some length (pp. 83-86, *Beekeeping for All*) the traditional method of populating a hive by driving bees from one hive to another, both usually skeps, by drumming with sticks.[14] Although this can be applied to frame hives, it is not popular, so it will not be covered here.

6.2.3.2 Nuc transfer by chop 'n crop – a method of last resort

This is a procedure used by top-bar beekeepers who buy frame nucs. I have not tried it myself, but Raimund Henneken, on whose report this section is based, found it quick and efficient. The bees are hived in a single operation, unlike in the growing down method described above.

The aim is to cut away enough of each nuc frame and comb to leave combs on top-bars that fit a Warré hive. This is severely intrusive as it involves cutting through brood comb, thereby killing larvae and pupae. So, in the short term, it is more stressful for the bees, and the beekeeper. Obviously, from the natural beekeeping point of view, it should be regarded as a last resort. However, it is no different in principle from doing a cut-out/rescue of a colony in a building. Having the comb already framed and on bars makes it all the more feasible than such a cut-out. It is advisable to have an assistant, preferably an experienced beekeeper who is informed in advance about the details of the plan. Visualise the entire procedure before attempting it, noting what additional equipment you will need, and even consider doing a dummy run. You will be doing 'cardiac surgery' on your colony! There is a serious risk of chilling the brood.

The nucleus may comprise five or six combs. If possible, find out from the supplier in advance the thickness of the top-bars at the point which will eventually rest in the Warré box rebate. When the nuc arrives, place it on the site that the Warré will occupy, with the entrance facing in the desired direction. It may be left there a few days if it has sufficient space for the queen to lay and is not already on the point of swarming. If the top-bars at the points to be chopped are deeper than the 10 mm (5/16") deep rebates in your Warré boxes, nail thin strips of wood all round the top rim of the Warré box to give the necessary depth.

There is no need to try to transfer any combs of honey. They are heavy and more prone to detach from the top-bar. Puff some smoke into the nuc. Search for the queen and put her inside a cage. (A clip type combined catcher and cage is particularly convenient. Fig. 6.5)

---

[14] For photos of driving see http://ruche.populaire.free.fr/essaim/tapotement/

Fig. 6.5 Spring-clip queen catcher/cage

Move the nuc aside, and replace it with a Warré hive (floor with entrance reducer, one Warré box with top-bars positioned, top-bar cloth and quilt). The advantage is that the field/foraging bees can immediately find the new hive and will not fly around throughout the job. Put the queen in her cage on the Warré floor.

Work on the nuc some metres away from the Warré hive, which is now collecting the field bees (returning foragers). Take the frames from the nuc, one at a time, spray both sides with a little water and shake the bees off into a swarm box, a simple box with a lid and openable vents covered with bee proof mesh. See § 6.3.2.1 for shaking technique. If there is a lot of uncapped brood, be careful when shaking the bees off the combs as the larvae could fall out. If necessary, remove the bees from the combs by brushing. Store every bee-free comb in a box some metres away from the swarm box. Put the lid on the swarm box, open its vents and put it in the shade near the Warré hive.

Take the box with the frames and go to a previously prepared place to chop and crop. Work on a surface which you can replace or easily clean afterwards! It is helpful to have a jig to show you where to chop the frame. (Such a jig must of course be constructed in advance.) Now saw through the frame and then cut the comb between any reinforcing wires with a knife, cutting upwards towards the top-bar. Finally use scissors to cut the reinforcing wires. Put the cut comb in a second Warré box and cover it to keep it warm. Instead of a saw, some beekeepers use tree-lopping shears to cut the frame's top-bar.

If you wish, you may recycle the side fragments of comb removed during cropping. Prepare in advance a special frame to support these fragments. If possible, ensure that the comb cells are correctly orientated.

Put the Warré box filled with combs on top of the first box on the original site of the nuc. Remove two or three top-bars at an edge of the box to facilitate transferring the bees. Pour the bees from the swarm box back into the hive. Replace the top-bars. Release the caged queen into the hive and cover the hive in the normal way. If forage conditions are poor add a feeder, especially as you may have removed most of the honey stores. Any honeycomb not left on the top bars can be fed back to the bees in a top feeder after first uncapping it.

Clean the surface where the comb was cut and any residues of comb or honey about the place so as to avoid robbing breaking out.

Fig. 6.6 Chopped and cropped frame comb for transfer of a nuc. Photo: Raimund Henneken

7   General management and monitoring progress

Subject only to time and distance constraints, how often you visit your hives is a matter of personal choice. Warré appeared to believe that the hive could be managed by only one or two interventions a year. However, if you are concerned about your bees, as beginners usually are, you might find it hard to stay away from them for more than a day or two. Obviously, the more frequent the visits, the sooner can be the remedy if a problem arises. Careful observation of the phenomena – keeping records or a diary, if you wish – eventually metamorphoses into a 'feel' for whether all is well at the apiary. On arriving at the apiary in spring and summer, my first task is to look round for swarms.

Storch details how a lot can be learnt from entrance activity.[15] All is well if, on warm, rainless days, the bees are coming and going purposefully, a good many carrying pollen. Before a brood cell is capped, the larva it contains needs a little under half a cell each of honey and pollen (bee bread) and about one and one third cells of water. Adult bees too need pollen and honey, the latter also supplying the energy for keeping the brood nest at 35°C (95°).

For some years, I used to assess entrance traffic by counting the number of bees returning per minute but my mentor regarded this counting as somewhat obsessive! The returning bees may be carrying nectar, pollen, water or propolis, or indeed be unladen after a flight for orientation, clearing or scouting. Above 240 bees per minute (20 in 5 seconds) is difficult to count accurately. You may also want to count pollen carriers to get an idea of what fraction they represent of the returning bees throughout the season. The entrance traffic of course depends on many factors but can give some idea of relative colony strengths.

The top box can fill with comb in a fortnight under UK conditions, sometimes less. In about another fortnight, though always ahead of demand, you may nadir a third box underneath (§ 8).

If there is no entrance activity, for example in winter, listen to the cluster by putting an ear to a box by the brood nest. A murmur/rustling/buzz shows that all is well. If the box walls are especially thick use a cheap stethoscope with the diaphragm removed and the tube pushed in the entrance. Accustom yourself to the normal smell of the ventilation draught at the entrance.

Note that for routine monitoring there is rarely any need to open the hive at the top, which would let the heat out. If the hive has no windows, viewing holes, nor floor modified to insert a mirror and light, and you have no lift, to see how far the nest has progressed and whether another box is needed, very gently slide the hive backwards on the floor about 50 mm (2") to make a opening. Take care not to crush any bees in the entrance. Use a torch and mirror to view the base of the nest, or, though perhaps more intrusively, take a flash photo. An alternative when the hive comprises only two boxes is to remove the roof and quilt, slide the hive about a finger breadth forward on the floor and, standing at the side of the hive, tilt it gently backwards while supporting it firmly at the back and

[15] Storch, H. (1985) *At the Hive Entrance. Observation Handbook: How to Know what happens inside the hive by observation of the outside.* Transl. from *Am Flugloch*, European Apicultural Editions, Brussels.

Fig. 7.1 Different colonies at various stages of development in the bottom box

look underneath. This can be done with even three well-propolised boxes, but be sure you have the strength to steady the hive while looking! Here is an opportunity to inspect and remove (or sample) floor debris, if any. Open your hive *rarely* so as not to unduly agitate your bees. Smoke is not normally needed for such inspections.

A little more invasive inspection involves peeling back the rear of the top-bar cloth part way. Reassuring signs are plenty of capped honey by the top-bars, and bees coming up to investigate.

If for any reason you need to inspect the brood nest, an intervention that Warré beekeepers generally like to avoid, it can be done on the whole box basis or by inspecting individual combs (§ 7.2).

7.1 Occasional phenomena in front of the hive

Orientation flights: Bees destined to become new foragers will be seen orientating at the front of the hive and spiralling further away to get a picture of its situation. When these events are more frenzied, beginners sometimes think that a swarm is issuing. Sometimes this may not be too far from the truth. Jürgen Tautz has presented evidence that such 'mass orientation flights' are preparatory to nuptial flights of young queens, even if on a particular day she does not emerge. With patience, you may even spot a queen leaving or returning, sometimes with the mating sign trailing behind. Her wing buzz is different from that of workers.

Piping: I kept bees for a few years before I heard my first piping of young queens.

This haunting and surprisingly loud noise, a combination of a reedy piping and almost a 'quacking', is easier to hear in the quiet of a windless evening in the spring/summer.

Corpses: Examining ejected larvae/pupae and dead or dying bees could help with disease diagnosis. I have a slab in front of most of my hives which helps with such monitoring.

Drones: Eventually, on a warm day in spring, the first drone flights take place. In the late summer, though sometimes later, the drones are forcibly ejected from the hive and many killed.

Guards: Sometimes a group of guards can be seen examining a suspect intruder bee, which stands in an appeasing posture throughout the process.

Robbing: Generally wasps or other bees only seriously rob a hive when it is in a weak condition, e.g. when queenless, so persistant robbing warrants closer inspection. Robbing is most likely to happen late in the season when there is little or no nectar available. Bees are seen constantly probing cracks round hives and testing entrance defences. If you notice robbing is occurring, e.g. many wasps entering, or bees with shiny abdomens leaving, or small particles of comb at the entrance, try fitting a robber guard (§ 2.4). I have witnessed two cases of a robbing war breaking out in my apiary, both times, as it happened, resulting from a visit by an inspector. Order was restored by restricting entrances (See Fig. 7.4) and leaning glass sheets against them. The chicanes created were no obstacle to the rightful occupants, but restricted access by robbers.

7.2 Brood nest inspection

For various reasons, and then only rarely, the hive will have to be opened at the top. You will have with you at least the basic tool kit including a lit smoker.

7.2.1 Inspecting boxes

Remove the roof and quilt but not the top-bar cloth. Free the top box and lower boxes

Fig. 7.2 Warré box stand

as described in § 8.1. If you are interested only in the top box, slide it about 25 mm (1") forwards, rotate it in a vertical plane 90 degrees onto its back and slide it forwards to rest on the box below, letting the bottom rim of the box sweep bees out of the way as you slide it, if necessary smoking bees clear too. You may now inspect between the brood combs, parting them gently with the fingers while lightly smoking bees out of the way. From the capped brood that is visible, you will easily be able to judge the extent of the brood nest. To check for larvae, you may need to part the combs further and look at an angle. In which case, be sure to smoke the bees towards the top-bars. The queen might still be among them and at risk of being damaged. It has been suggested that a dentist's mirror would be useful for this examination. If you need to look at lower boxes, proceed in the same way, putting each box inspected aside on a Warré box stand. If lower boxes are temporarily put aside, cover them with a spare top-bar cloth to retain warmth.

If necessary, with an assistant, the whole hive could be placed on a table frame, i.e. with the tabletop removed, and inspected from underneath.

7.2.2 Inspecting combs

Inspection of individual cells in the comb is possible with the Warré hive. It is normally done only on the order of an official inspection or if the beekeeper is undertaking his own monitoring for clinical foulbrood. As already mentioned, it requires greater care than removing framed comb. If you are in a country/state that requires comb to be quickly removable, you may need to consider using a frame or semi-frame version of the Warré hive. For those in the UK and in some states of the USA, if you feel that your inspector is unacceptably rough in his work, you are within your rights to remove the combs for inspection, and replace them yourself. Indeed, this could speed up the inspector's work: you remove/replace comb and he or she studies the brood. However, consider that many inspectors may have far more sensitivity for bees than the novice, and go through a box in exemplary fashion. Several UK inspectors already have their own L-shaped comb knives and are used to removing combs from Warré hives.

Unless the combs have been built parallel to one another, and do not cross between top bars, it is impossible to remove them without doing a lot of damage. Firstly free the box being inspected as described in § 8.1. The comb is usually fixed to the walls at each end at least half way down from the top of the box. This attachment has to be cut. Although it could be done from underneath with a long serrated knife, after first tilting the box on one side, an easier solution is to do it from above using the 'L'-shaped comb knife (§ 4.2).

The comb knife is inserted downwards between chosen combs with the shaft touching the inside of the box and the blade pointing away from the box wall. Insertion is stopped at a pre-marked distance where the knife is very slightly above the top bars of the box beneath (210 mm, 8¼"). The knife is rotated 90 degrees so that the blade goes under the comb to be cut. It is then pulled upwards, cutting close to the box side and freeing the comb. It is important that the cutting is always towards the top-bar that the comb is on, avoiding any downward force on the comb that might lead to detaching it from the bar.

When the knife reaches the underside of the top bar, the knife is further rotated, moving the end of the shaft away from the box wall as the blade is withdrawn, ultimately orientated parallel to the comb. The knife is then lifted out with the blade now pointing in the direction opposite to the one when it was inserted. The blade undergoes a 180 degree rotation in the process of cutting one end of a comb. This procedure only minimally disrupts comb cells and causes little seepage of honey.

Once the comb is free from the sides, you must now free the top-bar if fixed. As mine are pinned, I ease up each side of each end a little at a time with the help of the hive tool until the bar is clear of the pin or has come out with it. To lift the bar vertically, I use two hive tools at the same time, one each side of the bar. Alternatively, use the 'J'-type tool, resting its notch on a top-bar or on the hive wall.

Finally, place the comb holder on the top of the hive, gently lift the comb out and

Fig. 7.3 Warré comb removed for inspection; note some cells glistening with nectar.
Photo: Karman Csaba (see also back cover)

set it on the holder. Inspect the cells, if necessary with the help of a torch (flashlight).

Replace combs in the original order and orientation with respect to the entrance. Initially, they may be replaced loosely a little to one side of the box to facilitate working on later combs. The spacing for their final repositioning is usually very clearly marked by the deposition of propolis in the rebates. If all the combs have been removed from a box, for example when making an artificial swarm, they may be quickly repositioned using a top-bar spacing template, assuming you have made one.

7.3 Ventilation

A factor to be aware of during a hot spring or summer is whether there is sufficient ventilation. In cool temperate climates, such as in northern France where Warré lived, an entrance area of 18 cm^2 (2¾ sq.in.) is adequate for most summer conditions. But in hotter climates a wider entrance or ventilation higher up the hive, might be needed. Overheating in the hive is evidenced by 'bearding' round the entrance and under the alighting board. A proportion of the bees come outside in order to reduce heat production inside and improve ventilation. Sometimes it is a temporary phenomenon during the hottest part of the day, but it can also sometimes last all night. If ventilation is a concern, a wider entrance may be created temporarily by sliding the hive forwards on its floor to open up a full width gap between the rim of the bottom box and the front edge of the floor. Note that the hive will be less stable in this situation. Cooling at the top may be created by removing the quilt contents, and if they are in a cloth bag (e.g. old pillowcase) this is quickly done. Alternatively rest the quilt on four small blocks of wood so that its contents retention cloth clears the top-bar cloth. In this situation, you may be lucky to repeat Warré's observation that the bees sometimes unpropolise holes in the weave of a sacking top-bar cloth. In situations where the hive does not get enough shade in the heat of the afternoon, you may need to provide ventilation higher up, between boxes. The top box may be prised up a corner at a time and matchsticks or thin twigs inserted in order to create a ventilation slit without giving access to bees or wasps. In my climate in Wales, none of these extra ventilation measures have been necessary.

7.4 Stores

Dire conditions must prevail for a Warré colony to be short of honey stores in the spring or summer seasons. If the weather was poor after you hived your bees, you may have given them a feed for a few days or so. And if there is a long rainy period preventing the bees from foraging, you may need to feed again (§ 11). If you are concerned that stores are short you should check. If you have boxes with windows, look at the top of the comb in the top box, using a light if necessary. You should see capped honey higher up, especially at the sides. If you do not have windows, the inspection may be done with or without opening the hive. To check by opening, have a lit smoker at the ready, remove the roof and quilt, peel back the top-bar cloth about half way, starting from the back and examine the tops of the combs. There should be capped honey, especially at the sides. Use the smoker to drive the bees down if the combs are too crowded to see. Replace the top-bar cloth, if necessary smoking bees down as you do so. To check without opening, heft the hive or top-box (§ 9.3).

7.5 Slow growth

A colony may develop frustratingly slowly. Causes include: long periods of rainy weather, including very dull skies if not actually raining; drought; low temperatures; poor nectar and/or pollen availability; too few bees to start with; a poorly mated, old, diseased, and/or genetically feeble queen; high pest burdens in the bees and brood; and diseased brood. In the case of imminent starvation, you would most likely want to feed your bees. But if nutrition is not at crisis point, consider also letting the bees cope with dearths.

A number of Warré beekeepers have noticed that a colony may fill one or more boxes down to the top-bars of the next box, then stop, or even swarm through lack of space. This has come to be known as the 'false floor effect' in that the top-bars appear to act as a hindrance to colony development. Contributing factors could be adverse weather and forage conditions. I let my colonies develop at their own speed and even allow them to swarm. For the impatient beekeeper, or one who, through local or other circumstances, must try to avoid swarming, a possible, though invasive, remedy is to move one brood comb down into the empty box to help draw the workers' attention to the extra space. I and others have experimented with putting the top-bars on edge in order to create wider gaps between them. The result was that the bees built their combs through the gaps making the hive unmanageable. The only other way to get round the problem, although equally inconveniently for the beekeeper, is to do away with top-bars in the lower boxes and substitute a couple of spales in each box, as is done in Japan with *Apis cerana* beekeeping in hives nadired like the People's Hive.

7.6 Honey binding

Two reports have reached me indicating that productive Warré hives have been so inundated with nectar that the brood nest has been constricted to a narrow cylinder down the middle of several boxes. In Canada, this 'honey binding' has resulted in increased winter losses, presumably because the geometry of the hive and cluster was far from what was thermally ideal. The bees had too little brood cell space in which to form a tight spherical cluster, thereby minimising the surface to volume ratio and consequently reducing heat loss. Part of the solution adopted by the beekeepers concerned was to super the colonies in the main flow, i.e. add boxes at the top of the hive. This may be done without putting a queen excluder on, as the depth of honey in the top box itself acts as a reasonably effective queen excluder. The supers could be empty Warré boxes with top-bars, with or without a 'ladder' of drawn comb or foundation fixed to a middle top-bar, or a box of ready drawn comb. In the former case, most of the comb is likely to grow up from the top-bars of the box on which the super is set. It will therefore be much more difficult to harvest, involving a good deal of honey leakage and risking drowning bees. In which case a far slower, more cautious, removal could be effected by allowing the bees to mop up leaked honey as the box is lifted slightly while drawing a cheese-wire through the comb connections.

7.7 Defective queen

In spring and summer, suspect a failing queen if, having excluded poor foraging conditions, there is low entrance traffic, little pollen coming in, and, in due course, lots of drones at the entrance. Having another hive (or more) is useful for comparison. Early remediation, i.e. when there are still (potential) nurse bees, has a better chance of success, so consider inspecting the brood. If you see hardly any worker brood – only patches on a comb or two – together with no occupied or hatched queen cells, or worse still, the domed cappings of drone brood in worker cells, you may wish to requeen, especially if you have invested a lot in your first and only hive. See § 7.2 for methods of inspecting comb in situ or by removing it. Your supplier of bees will normally be able to provide a replacement queen, in the case of a recently hived package, either free or at a relatively low cost. If you work only with swarms and need to requeen, you may prefer to requeen by uniting your colony with another swarm or cast (§ 7.8). No requeening method is infallible. Least success is had with introducing a queen who is not already accepted by workers. Requeening with a mated queen saves about a month's delay in egg-laying that would otherwise occur if you merely introduced a comb of eggs, or, if eggs are already present, left the colony to make emergency queens.

Before introducing a new queen you must remove the defective one. First, you must find the queen, either by inspecting individual combs (see above) and temporarily putting them aside in a spare box as each is inspected, or, if you have no success that way, by the far more disruptive filter method of last resort using a queen excluder. In order to avoid panicking the queen and making her harder to find, use no smoke when opening or searching comb by comb. For the filter method, Warré writes: 'Put all the occupied boxes of the hive on one side. On the hive floor place one or two empty boxes according to the strength of the colony. On top of the empty boxes place a queen excluder. On top of the queen excluder place all the boxes of the hive previously put to one side. Open the top box and smoke strongly and rapidly between the frames. ... When the bees have abandoned the top box, proceed in the same way with the other boxes. When the queen excluder is exposed, the queen will be found amongst several drones. She is destroyed, or, if she is going to be used, placed in a cage.' Rather than discard a failed queen, you may wish to preserve her intact in alcohol and build up a collection of queens to make a pheromone lure, in which case kill her by placing her in a freezer.

If during inspection you notice one or more occupied queen cells (larvae in royal jelly or capped), the slow developing colony is in the process of superseding its queen. In this case you may wish to leave the colony to requeen itself naturally. But if there is little or no brood at all, despite forage being available, you may wish to proceed to requeening.

Between about one and two hours after making the colony queenless, the new queen is introduced in a queen introduction cage so as to accustom the bees to the impostor's aroma. The travelling cage in which she is supplied may adapt as an introduction cage, in which case remove the attendants and expose the candy plugging the releasing hole. This is done in a transparent plastic bag. Alternatively, work in a car with the windscreen vents covered, or in a closed room and catch the queen when the bees fly to the window. If you have to transfer the queen to a separate introduction cage, take the same precautions to retain her. Place the cage between the combs at the top of the brood nest, near capped brood if there is any, making sure that the candy hole has candy in it, the plug/tab over the candy is removed, and that neither the mesh nor the hole are obstructed by comb, hive wall or top-bars. Check it in a few days. If bees are still apparently attacking it, leave it for longer. When the queen has been released from the cage by the bees chewing away the candy, remove the cage. Thereafter leave the hive severely alone. It will be at least 21 days before your colony population resumes its build-up.

7.8 Weak colonies – uniting

If a colony is dwindling, appears disease free, and you do not want to attempt to requeen it by the introduction cage method, then the option is to save the bees and comb by uniting it with another colony which shows more promise. This may be either a swarm or a colony already established in a Warré hive.

Commonly the advice is first to remove whichever queen you do not want. However, one may also let the bees of the united colony decide for themselves which queen to keep. In my experience, there is almost negligible risk that the better of the two, or both, will be killed. I usually unite over newspaper, giving the main colony control of the entrance. Fold a page of newspaper and perforate many times with the point of a knife to leave little slits but not openings. I do this on a thick carpet. Take the paper, water spray and basic tool kit to the hives to be united. Remove the roofs and quilts. Remove the top-bar cloth of the receiving colony that will have control of the entrance at the end of the operation, and replace the cloth with the newspaper. Thoroughly wet the paper. In windy conditions, temporarily weight it with sticks or pebbles while spraying. Place the occupied boxes of the hive to be united on the newspaper and replace the quilt and roof. Leave for at least 24 hours. The bees chew through the paper, colony smells mingle and the bees accept each other. Paper shreds at the entrance show that all is going well. If the resulting tall hive is to be reduced, for example for wintering on two boxes, smoke the bees down out of any unwanted boxes of the top hive and remove the boxes. Any broodless boxes to be retained are best placed below the existing brood nest. Boxes containing a lot of honey may stay on top.

With uniting swarms to an existing colony, for example for requeening, the method is essentially similar, except that first the swarm is hived in a Warré box and then placed on top of the receiving colony. If the swarm is small, it may take longer than a day for the paper to be chewed through and proper acceptance to be completed. Only occasionally does fighting break out with this method of uniting.

A 'quick and dirty' method of uniting a swarm avoids initially hiving it in a separate hive. An empty box serving as a funnel is placed on the receiving colony with its top-bar cloth removed and the bees smoked generously from above and below (through the entrance) to disguise scents. The swarm is then dumped into the empty box and the bees smoked down, When the empty box is largely clear of bees, remove it and smoke the bees down between the top-bars. Replace the top-bar cloth, quilt and roof. There is a somewhat higher risk of fighting this way. Instead of smoke, a gentler method is to use a spray of sugar syrup (1: 2 sugar:water by volume) scented with a drop of some essential oil, e.g. peppermint, to douse both the host bees and the new arrivals. As with the smoke, it masks the different scents of the two colonies in the ensuing chaos after mixing, and the activity of cleaning off the syrup spray distracts the bees.

7.9 Swarming

Colonies are unlikely to swarm in the same season that they are hived. But it does occasionally happen. A really good queen combined with the long availability of good forage can trigger the issue of a late prime swarm. So it is as well to ensure you always have space to build below the brood nest so as to rule out, here at least, constriction through lack of space. If when nadiring you notice a thick cluster of bees hanging below the combs and apparently doing nothing, this could be the bulk of a swarm due to issue in the ensuing days. Bearding at the entrance on a hot day is a sign that this cluster has temporarily moved out of the hive to help reduce the temperature inside. Activity at your bait hive entrance(s) is another indicator of the possibility that a swarm may be considering issuing. For more on swarming see § 14.1.

Fig. 7.4 Temporary mesh robber guard over entrance. Bees bringing in ivy (*Hedera helix*) pollen; bee coated with Himalayan balsam (*Impatiens glandulifera*) pollen on right.

8   Nadiring – adding boxes underneath

I add a third box when the second is about half built. Occasionally, I have been caught out and arrived to find the second box full to the floor. The snag then is that the shortage of space may already have triggered the swarm mood. Also, when the brood nest is right to the floor, the colony may be a little more defensive than when the comb ends a few inches higher up. Warré states that all the boxes needed for a whole season could be added at the spring visit, say a little before the dandelion flow. That may be fine for northern France, where perhaps four or five boxes would be the maximum required, but in more melliferous localities it would be impractical in that it would increase the risk of the hive blowing over through its low initial weight relative to its height.

There will eventually come a time when even a 2-box hive is too heavy for one person to lift, for example when nadiring a third box. With a suitably protected assistant present, remove the roof and quilt, free any adhesions round the floor with the hive tool, and, with an operator standing each side, lift the hive forwards onto a Warré box stand. At this point, you may wish to inspect and scrape away any floor debris, or substitute a new floor. Add the new box with top-bars in place, put the hive on top, maintaining its original orientation, and close up.

If neither assistance nor a Warré hive lift are available, then there are two options. The first, dismantling and reassembling the hive a box at a time, is highly invasive, but may be the only option you feel you can safely manage (§ 8.1). The second, reported by John Moerschbacher, is less invasive for the bees but more strenuous for the beekeeper, involving a series of steps upwards as the hive is moved back and forth onto supports (§ 8.2).

8.1 Removing boxes

Any hive manipulation involving separating and replacing boxes runs the risk of squashing bees. Apart from the fact that you do not want to lose bees, it is worth keeping in mind that a squashed bee can trigger alarm in the colony and thus make your job more difficult. For such manipulations, a feather and, if necessary, the smoker are your tools for ensuring that surfaces such as floors and box rims are free of bees before closure of any gaps. Slowly sliding boxes into place is usually safer than lowering them. But even with sliding, you need to watch out for bees becoming trapped as gaps between rim and rim or floor and rim close up.

A further issue is the risk of provoking the bees into balling the queen thereby damaging her with their mandibles. In the worst case this could result in her death by her overheating or being stung. While manipulating Warré hives, I have not yet seen balling, but I have very occasionally seen it with my National frame hives. The risk is higher with newly introduced queens. A remedy is to smoke the ball to disperse the bees and close up the colony as soon as possible, if necessary postponing the manipulation that was to be done. This phenomenon should be kept in mind in all operations involving opening the brood nest.

On a sunny, warm day when the bees are busy foraging, have ready your basic tool kit including a lit smoker, wedges, box stands, clean floor (optional) and your empty box ready with waxed, positioned top-bars. Warré advises puffing smoke in at the entrance to tell the bees you're there. I never do this routinely. If I use smoke it is only when required to control the bees. This manipulation will very likely call for smoke.

Remove the roof and quilt but leave the top-bar cloth in place. Using the hive tool, gently prise up the top box from the box below all round lifting no more than 3 mm ($^1/_8$") as you go. This gap is too narrow for the bees to exit. Slowly rotate the box a few degrees clockwise and anticlockwise to break any comb bridges to the box below. If it does not rotate with ease, do not force it. You need to cut the bridges with cheese-wire.

The safest way to do it as regards not garrotting bees is as follows: starting from the front, lift one corner of the box with a hive tool placed 25-50 mm (1-2") from the corner, so that you can get the wire under the corner. Do the same with the other corner. Once the wire is in, insert a wedge at each corner behind the wire's direction of travel so that they support the weight of the box at the front with about a 3 mm ($^1/_8$") gap, thus avoiding trapping the wire. Now you have the wire under both corners and the corners resting on wedges. Gently and slowly pull the wire towards the back, one side first, then the other, then the first side and so on. Always cut along the combs, moving from the front to the back, never parallel to the combs! A slight sawing motion may help. When the wire is about 30 mm (1¼") from the rear outer wall the danger arises of scissoring bees against the edge of the inner wall. Rather than completing the cut through the seam, stop at this point and remove the wire slowly in the reverse direction, finally removing the wedges.

Any bridges remaining at the back may be broken by tilting the box backwards. At this point you really are opening the hive, and are right in the heart of the nest, so it is as well to proceed cautiously. With the hive tool and smoker at the ready, lift the box a little and support it on the thick end of a wedge or on a piece of tree branch. Watch for a while how the bees respond to the light and cold air being let in from an unfamiliar angle. If they seem fairly calm, you may continue. If they seem agitated, gently puff smoke into the seam. If they attack your veil, consider that it may not be the right time to open up fully. A colony's mood can change from one day to another. If you decide to close the hive, to avoid squashing bees in the gap, smoke them back from the box rims. But if no other time is available to you, more liberal use of smoke, combined with rapid though not violent working, should see you through.

Whenever top-bars are exposed in manipulating a hive, Warré advises scraping them free of comb and propolis with the hive tool. I never do this as a matter of routine. Not only does the vibration annoy the bees on the combs below, but also it seems largely unnecessary. The only time I feel the need to scrape is if more than a thin layer of comb from the box above has been left on a top-bar of the box below.

Complete lifting off the box and set it aside on a Warré box stand (Fig. 7.2). Repeat with the next box and so on until you reach the floor. As you progress, cover each box with a spare top-bar cloth, or piece of cardboard, etc., in order to keep the heat in. Use your smoker if the bees become agitated. If the floor needs scraping clean, smoke or brush any bees off it first. You may wish to inspect the floor for varroa or other diagnostic signs. In which case, to minimise the length of time the hive is open, substitute a clean floor. Place your new box on the floor and rebuild the hive taking care to maintain the former orientation of boxes and avoiding crushing bees between box rims. The smoker and brush help greatly with this. Alternatively the box may be slid into position from the back or front, slowly sweeping before it any bees on the rims and top bars of the box below. Caution near the end to avoid guillotining bees!

8.2 The stepwise method of nadiring

According to John Moerschbacher, if this procedure is done quietly and slowly, one hardly disturbs the bees at all, especially if they have not yet built comb to the bottom of the bottom box. However, on account of the several moves, as there is a higher risk of inadvertently jolting the hive, some may wish to hold the method in reserve as a last resort. Have a lit smoker ready. You may not need it to calm or control the bees, but it could be useful to clear bees from surfaces to avoid crushing them.

Nadiring this way is easier if the hive is close to the ground, for example if the hive floor (bottom board) is on bricks or pieces of thick wood. The method can be done in two or more steps. Taking more steps gives greater stability during the move. The method relies on the boxes being propolised together. However, one should not put undue strain on the joints. If in doubt whether the boxes will stay in place during the move, install hive clips/springs to lock the boxes together. This method does not give access to the entire surface of the floor for scraping it clean.

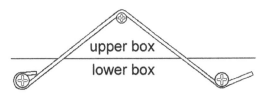

Fig. 8.1 Z-spring clip hive fastener

Put a temporary base a few inches higher than the floor just behind the hive: e.g. bricks, an old feeder, a hive box, anything. Kneeling in front of the hive, tilt it forward, and slide it back across the floor so that the back rim is on the temporary base. The front end is still resting on the floor. Put the box to be nadired complete with top-bars on the floor, forwards of its final place. Tilt the hive back far enough to bring the bottom of the front just above the top of the new box and slide the hive forwards to rest its front on the new box. Standing behind the hive, lift the back end and slide it into place on the box. Slide the entire hive back into its final position on the floor. Reposition the floor if it was displaced during the move.

For those who feel less confident about supporting the hive with the degree of tilt that this involves, especially when there are three or four boxes, doing it in more than two steps may help. For example, each step would be a quarter rather than half a box height, all executed with suitable pieces of wood or hive components (quilt frame, ekes, etc.).

8.3 Mechanical lifting

Apart from the effort of getting the lift to the apiary – not an insignificant effort with a lift the weight of mine – this is by far the most comfortable way of nadiring new boxes. Although I have a smoker with me I hardly ever light it for this job. I remove the roof, adjust the lift forks to the desired height, slide the lift in on top of the legs of my hive stand with the forks under the handles of the box to be lifted, apply a little upward force to the box by turning the lift crank, free the joint to be separated with the hive tool and lift the box to the desired height. A new box may be slowly slid in from the front, sweeping bees before it, taking care not to guillotine any bees as the box comes to its final position. If there are so many bees in the way that you risk squashing some of them, use smoke to clear them from the box rims. The hive is then lowered onto the new box, if necessary, as you lower, clearing bees from the box rim with a feather, or with the smoker, if lit.

Fig. 8.2 Nadiring a new box with a lift

9 Harvesting

Here I cover harvesting honey and comb. Wax is a by-product of harvesting honey, but I prefer to leave to the bees their other hive products such as royal jelly or propolis. Harvesting comb from the top of the Warré hive is part of the comb-renewal process built-in to its management.

If there is a genuine surplus of honey, i.e. more than the bees need to survive, you may harvest by the comb (§ 9.4) or by the whole box. Warré advises having only one harvest and this at the end of summer, or in early autumn (fall). There has been some discussion as to whether this is always ideal, some arguing that it is better to leave all the honey for the winter to avoid the risk of removing too much at a critical time, and then harvest in spring. But here too one could risk leaving the bees short over the summer, if the weather turns bad, as it has done over six seasons in my locality. I harvest in September when most of my main flows are finished, but there is normally a steady nectar input until November from, initially Himalayan balsam (*Impatiens glandulifera*), Japanese knotweed (*Reynoutria japonica*), and, ultimately ivy (*Hedera helix*). I do not rely on these flows, but in September I determine how much honey the bees have and what I can safely take. If there is a genuine surplus, i.e. more honey than the bees need to survive, you may harvest by the comb or by the whole box. This might not occur in your first season, especially if you started the colony late in the season. However, it is not possible to make a hard and fast rule about this, because, in some places, colonies gain spectacularly in weight.

Honeys such as oilseed rape (canola) and ivy that set quickly in supers above queen excluders on frame hives, have not proved to be a problem in the Warré hive, provided that the honey is harvested in the season it is collected by the bees and extracted within a day or two of harvest. Whether the apparent lack of problems with these honeys is due to the more even distribution of warmth in the hive, possibly because of the freedom of the queen to roam, as has been proposed, is still a matter of speculation.

Honey contains organic acids which corrode some metals. When processing and storing honey, the following materials are acceptable: glass, glazed ceramic, stainless steel, and food grade plastic.

9.1 How much surplus honey?

In the last few years, from several of my hives the harvest has been nothing at all, or at most a comb's worth, whereas in other countries Warré beekeepers have reported harvests of up to 43 kg (95 lb) per colony (averaged over nine colonies). My last summer of high honey yield was 2006, with 20 kg (45lb) per colony, but that was before I started Warré beekeeping. Average honey yields per frame hive in the UK are said to be in the range 14-18 kg (30-40lb) and a regularly conducted honey yield survey in south-east England showed an average of 18 kg (40 lb) per colony over the years 2005-2012. For 2010, the United States Department of Agriculture reported average honey yields per colony of 30 kg (66lb). Warré hives are expected to produce less than frame hives because there is generally no supering and none of the re-use of comb associated with supering. Jean-Claude Guillaume, who was in Belgium at the time, did a cost-benefit comparison of Dadants (frame hives) and Warrés, allowing an average yield of 12 kg (26lb) per annum for his Warrés.[16] When there is no supering of re-used comb, more nectar goes into beeswax. This is valuable and is of course eventually rendered and purified for sale, so should be factored into the yield statistics. In any case many Warré beekeepers are not particularly concerned about getting honey from their hives, regarding it as an occasional added bonus to the pleasure of keeping bees.

9.2 How much can I take?

When harvesting at the recommended season, leave enough honey for winter. Warré in northern France stipulated 12 kg (26lb), which, in my experience, is at *least* six combs completely full of honey. In mild, maritime, west Wales I find that 9 kg (20lb) is usually sufficient. This is about three quarters of what I leave on my frame hives. Steer on the safe side, at least until you know what is required in your locality. Warré beekeepers live in greatly differing climates. You could find out from local beekeepers what average weight of honey their frame colonies need, and, as Warré found that frame hives needed more honey to winter on than the People's Hive, you could aim for about 80% of that average.

To count stores by the comb, open the hive as described in § 8.1 and count combs of capped honey from *underneath*. This is important because viewing from above will not show whether combs are full. As this method is disruptive to the colony, and you may not need to open the hive at all if there is no surplus, the other way of assessing stores is to heft the hive.

9.3 Hefting

Experienced beekeepers heft their hives by lifting the back, or in the case of the People's Hive, one side, and 'feeling' the weight. You could develop this faculty by loading boxes with different weights, say 5, 10 and 15 kg (10, 20, 30 lb), by using bricks and seeing how the boxes feel when lifted a little by a handle. To be more accurate, you may wish to use a weigher (§ 4.2). In my locality, it suffices to weigh only the top-box. In very cool climates where more than a full box of honey may have to be left, weigh the top two boxes. To do this, free the propolis seal round the seam to be broken and free the box from comb bridge to the next box as described in § 8.1. No bees are let out by this procedure. Hook the weigher under a handle. Lift the box no more than 3 mm ($^1/_8$") and record the weight. Do the same with the other handle. Add the two weights. Subtract the average weight of the box(es) that you recorded earlier (§ 4.2), the weight of the combs (about 1 kg (2¼lb) for 8 combs) and a 'guesstimate' for the bees of 1 kg (2¼lb) per box. If you are hefting in late winter to check if there are sufficient stores, you may allow only 0.5 kg (1lb) per box for bees. You should now have an estimate for the combined

---
[16] http://warre.biobees.com/guillaume_cost_benefit_analysis_2012.pdf

weight of honey and pollen. Be on the safe side by assuming that about a tenth of the resulting weight is pollen.

Less disruptive to the brood nest is to weigh two or more boxes without separating them. This might give slightly less accuracy. However, if there are plenty of stores, the top-box could then be weighed.

9.4 Harvesting individual honey combs

If there is only enough surplus to allow you a comb or two, it would be better to leave the harvest until the following spring and see how things stand then. This is because removing combs completely at this time will very likely leave a void in the nest and possibly impair its thermal efficiency in protecting the bees from the winter cold. Of course, you could extract the honey from the comb and return the comb, still on its top-bar, to the hive. Extracting the comb intact would require special cages that fit in an extractor and support the comb during centrifugation (p. 64, *Beekeeping for All*), but I know of nobody who does it this way.

If there is still a good honey surplus in the main spring flows, you could take your comb or two then, usually from the outermost bars. Removing combs with the help of a comb knife is described in § 7.2.2. Have to hand a lit smoker and the rest of your basic tool kit. In addition, you will need a food grade container with a tight fitting lid. This is placed right beside the hive with the lid loosened. For a few pence/cents, beekeepers in my locality obtain polypropylene buckets in which ingredients are shipped to local bakers and confectioners. Two hooks made from pieces of stiff wire in a similar way to the comb knife, only without the blade being flattened and sharpened, are useful to have to hand for use in the event of the heavy honey comb 'unzipping' from the top-bar. Use the hooks, one in each hand, to lift up the comb. Before removing the comb, lightly smoke down the bees on each side. If there are only a few bees on it, smoking is unnecessary. You may wish to leave any freed combs in place for a few minutes to allow the bees to clean up any honey spilled from open cells. Take care not to force the freed comb when lifting. When it is up, rest the bottom of the comb on the top of the box and brush off any remaining bees with a feather. Lower the comb into the container, cut it off the top-bar leaving about 5 mm (¼"), replace the lid and return the bar to the hive. Be aware of the risk of robbing being triggered by honey spillages.

9.5 Harvesting whole boxes

In this case, whether you are in your first or later seasons, your assessment of stores by counting combs or hefting allows you one or more boxes of surplus honey for yourself. If, on the other hand, there is no surplus to harvest, you are faced with leaving the box in place until the next season, thus missing the opportunity to remove old comb from the system. Though this is not ideal, it is no major problem. Beekeeping writers have reported using comb up to 20 years old. My mentor had supers comb that was over 30 years old! However, in the following spring, you could force box removal in the way I describe in § 13.1.

Along with your basic tool kit, take to the apiary either plastic bags big enough to put harvested boxes in or sufficient trays to go under boxes, and boards to go on the tops of the boxes removed. This is not only to catch spilled honey but also to prevent robbing.

Before you can take your honey, you must clear bees from the box. Warré and Guillaume advise smoking the bees down into the box below. In my experience this can be slow and not always entirely successful. Furthermore, one risks tainting the honey with smoke and putting little particles of charcoal in it which show up conspicuously in the honey pot, unless fine filtration is used. It is also less bee-friendly than the other clearing methods described below.

Firstly, free the box as described in § 8.1. You may if you wish raise the box temporarily onto three or four 3 mm ($^1/_8$") wedges, and, if comb bridges had to be severed, leave it about a quarter of an hour for the bees to clean up any honey spilled. Peel back the top-bar cloth about half-way and puff plenty of smoke under the cloth and between the combs. The cloth helps retain the smoke and its resulting higher concentration induces a faster retreat by the bees. When most of the bees are out of the box, remove it from the hive, transfer the top-bar cloth to the next box, smoke/brush off any bees still in the box and either bag it or set it on a tray with a cover to exclude bees. Proceed the same way with any other boxes to be harvested.

Instead of smoke, some prefer to place on top of the box a fume board soaked with a bee repellent, e.g. bitter almond oil or Fischer's 'Bee Quick'. The board has a shallow rim below it and a cloth pad affixed, onto which is deposited the repellent. The sun's heat evaporates the repellent, thus enhancing the effect. I have not tried this on a hive. Indeed, it is conceivable that it transfers traces of repellent to the combs. However, the

Fig. 9.1 8-way bee escape

same could be said of smoke. As the bees like neither smoke nor fumes, you may prefer a clearer board method.

Two clearer-board methods of clearing bees from Warré hives are described here, the first with the clearer board on the hive, the second with it off the hive. Plastic bee escapes (6 or 8-way; lozenge, Porter, etc.) are obtainable from apicultural suppliers. An 8-way escape fits inside the internal footprint of a Warré box. It must be mounted on a rimmed board with a hole cut in the middle and with rims that give sufficient depth above (approx. 20 mm, ¾") to allow access for the bees, and below (30-50 mm, 1¼-2") for bees that have passed through the escape to cluster.

Free the box(es) to be harvested. As they should be almost entirely honey, the queen is unlikely to be in them. However, you may wish further to reduce the chance of her being there by removing the top-bar cloth and applying a little smoke between the combs. The ingress of light and smoke should drive her down. Set aside the box(es) to be cleared, using the Warré box stand, or, with the hive lift, open up a sufficient gap below them to accommodate the clearer board. Slide the clearer board into position and replace or lower the box(es). I generally leave clearer boards on overnight. It might take longer. If the/a queen is above the clearer the procedure will not work as it depends on bees leaving the box to seek the queen below.

To clear a box off the hive, Andrew Janiak has reported a method that works well for him. Its advantage over on-hive clearing is that the colony is disturbed only once. The box is placed beside the hive entrance on a tray and covered with a clearer board with holes in it and on which are positioned bee escapes comprising bee-proof mesh cones of metal or plastic with a hole in the top for bees to exit. Plastic or metal fly screen mesh suffices for the cones.

Leave for between two and four hours. Emerging bees soon scent 'home'. Any bees left in the box are usually house bees and drones who are reluctant to leave. These may be removed with a little smoke and/or a feather.

Fig. 9.2 Left: external clearer board. Right:detail of bee escape.Photo: Andrew Janiak

10 Extracting honey

As this is a book primarily for getting started, I will concentrate on the simplest form of extraction using mostly utensils commonly found in domestic kitchens. Much of your honey will be extracted from comb that once held brood. Some say that this honey is inferior to that collected in the supers above queen excluders in Langstrothian hives. The error in that view is clear from the following observations.

Bees clean the cells out scrupulously after a bee has hatched. The comb in a Warré is in a continuous process of renewal. It has normally gone through only a season or two of brood cycles before back-filling with honey occurs, and the comb is harvested, never to be returned to the hive. Combs in the supers of framed hives are used over and over again, sometimes for 30 years or more. Pesticides and other artificial toxic substances reside in them and the extent of their later release into honey is unknown. Framed comb brood cells back-filled with honey may be weeks in contact with honey, long enough for the honey to pick up extractables from the comb before it is transferred to supers. This also applies in frame hives to honey in transit up the nest after it has been temporarily parked at the bottom by house bees who receive it from foragers. Vertical top-bar hives – managed and harvested in the same way as Warré hives – have been in use for at least 500 years in Japan and yet the honey from them is prized for its flavour. The corresponding period in Europe is at least 200 years. Honey is harvested from brood comb by honey hunters from wild colonies and by beekeepers from skeps, logs, etc., and has been for millennia. Some say that the flavour and bouquet is enhanced by its being from brood comb. To the argument that brood comb has larval faeces in it, it is worth noting that several foods, e.g. white bait, are eaten complete with the organism's whole gut. There may be antibiotic and other microbe inhibitory substances in the walls of brood comb which could enhance the qualities of the honey, e.g. from propolis. Most real food is not sterile when eaten. I am told by an importer of tropical honeys that top-bar beekeepers in east Africa, when offered a taste of 'pure' honey from the UK, did not consider it to be real honey.

Honey gourmets say the best honey is that which is the least processed. This would be honey served in the comb, often sold as cut comb or 'sections'. It fetches a handsome premium compared with honey in the jar. The next least processed is honey that is drained under gravity from the comb after it is cut. Closely following that is honey extracted by crushing and straining. Honey extracted by centrifuge, a process not covered here, is exposed more thoroughly to the air, thereby impairing its subtle bouquet. But ultimately it is all a matter of taste.

I have already mentioned two 'problem' honeys, namely rape and ivy, but there is one other kind in that category, namely thixotropic (gel-like) honey. For example, heather honey, which, without special equipment, is impossible to extract from framed hive supers combs by centrifugation, is easily extractable by pressing. Because the gel is inclined to re-establish, draining solely under gravity does not work very well, even if the comb has first been crushed. An old tried and tested method is to put the crushed comb in a muslin bag and work it with the hands. Nowadays, for hygiene reasons, one might wear close fitting, honey-proof gloves. Alternatively, the bag is squeezed between two boards that are hinged together at one end and have handles at the other.

Fig 10.1 Warré honey combs. Photo: John Haverson

10.1 Cut and drain

Invert the harvested box on a drip tray and cut the combs free from the side with a serrated knife such as a bread knife. With the box upright, lift a comb out and place it on a board or tray. Cut off the top-bar about 5 mm (¼") into the comb and place the bar over a suitable container to drip dry of honey. Its residual comb is the bees' starter when the box is reused. Examine the comb for any areas that are pure white or nearly pure white. These comprise comb that has not had brood and is suitable for consuming as cut comb honey. Cut these off in chunks and place in containers. Apiculture supplies sell small plastic tubs for this purpose. One of my customers even asks for cut *brood* comb, having once enjoyed wild honey in Africa.

If there are any combs free of honey, you may wish to keep these for bait hives after having inspected the combs for residues of brood disease (§ 15.7). In case they have wax moth eggs already in them, it is advisable first to store them for 48 hours in a freezer, and thereafter in sealed plastic bags.

Now check if there are any areas where the cells of honey are not capped. It is possible that the honey in these is not fully ripened, i.e. has not been dried down below 20% moisture content. In which case it will not keep but rather ferment, even in a refrigerator. If it is only a small proportion of all the honey, say below 10%, you can go ahead and use it. If it is more than that, it is advisable to test it. Hold the comb in both hands and with one sharp shake try to eject the uncapped honey onto a tray. If it stays in place you may extract it as honey. If not, you may feed it and the rest of the runny honey back to your bees, in which case first cut out the comb that contains it from the rest.

For minimum disruption of the comb for drained honey, slice the remainder of it into strips, each time cutting with a sharp knife through the middle of adjacent rows of cells. Every now and then, scrape the strips and exuding honey into a strainer over a bowl. Repeat with all the other combs. Occasionally you may come to pollen,

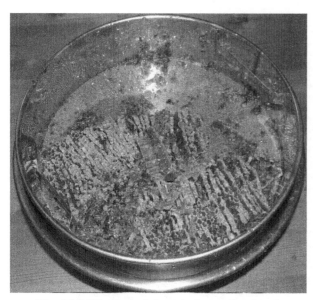

Fig. 10.2 Sliced comb in a strainer over a bowl

as bee bread, covered with capped honey. This may be distinguished from the rest by a different colour or thickness of comb. There is little point in adding this to the rest, so cut round it and remove it. You could consider storing it frozen for supplementary feeding to your bees.

Leave the comb to drain in a warm place for at least 24 hours. Some even put it in a car in the sunshine. Be sure that the temperature is not above 40°C (104°F), and there is no access to bees or they will come from miles around. Pour the honey into jars (§ 10.3) and store in a cool, shady place.

An ultra-flexible kitchen spatula is an ideal tool for getting the last drops of honey out of containers.

10.2 Crush and strain

For a quicker and larger-scale method, but one that might result in a cloudier honey due to release of comb fragments and pollen, chop the comb into chunks and place them in a container such as a bucket. Crush them with a suitable tool such as a potato masher or piece of wood. Strain the slurry either through a fine mesh kitchen strainer or through a bag designed for straining food, for example a jelly bag. Some use a paint straining bag. A large capacity and inexpensive straining setup can be made from such a bag and two plastic food buckets with lids. In the lid of the first bucket, the receiver, is cut a large hole to fit the base of the second bucket, the strainer. Several holes are drilled in the base of the strainer bucket and a filter bag is placed in it, folding the bag rim over the bucket walls. The slurry is poured in, covered loosely with the lid and the unit left in a warm, bee-proof place for as long as you like. Although this may produce a more turbid honey, some prefer more pollen in their honey. Indeed, I have customers who buy my honey specifically to administer to themselves the local pollen spectrum, believing that it reduces their hay fever.

If you produce so much honey that you have some to sell, you must comply with local food hygiene and labelling regulations. Your local beekeepers association may help you with this. You should have no difficulty selling it, either from your door, or through local retailers. Shops in my locality are always asking for local honey.

10.3 Bottling

When large amounts of honey, say 10 kg (22lb) or more, are to be bottled it can be a messy business trying to do this by ladling or pouring. The easiest way to bottle is from a ripener (settling tank). A ripener may easily and inexpensively be made from a food grade lidded plastic bucket of about 10 litres (2½ US gallons) capacity and a 40 mm (1½") plastic honey gate tap purchaseable from a beekeeping supplier. Using a pointed kitchen knife with a narrow blade, cut out a hole previously marked to the right size near the bottom of the bucket and fit the tap with the screw and seal provided. Before use, test the seal with a full bucket of water. Ensure the bucket is dry before putting honey in it.

The honey to be bottled is poured into the ripener and left for at least 24 hours for bubbles of air and particles of wax, propolis, etc., to rise to the surface. If the bucket is placed on a support of a suitable height above a table surface, the operator may sit and bottle in comfort. The lid is left only loosely on the bucket while bottling.

10.4 What to do with the comb

No matter how long you drain the comb, it will still contain trapped honey. You could feed this back to the bees (§ 11.2.2) or extract the last drops by pressing the comb fragments in a cloth bag (I use cotton) in a cider press or in a dedicated honey press bought from an apicultural supplier. Such a press, often called a heather press in the UK, is expensive but it is efficient, gives a relatively clear honey and reduces the comb to a dense cake ready for further processing to extract the valuable wax. Sausage stuffers are also used for pressing comb. Home-made presses are constructed from wood and the pressure is applied with a car scissor-jack.[17]

There are several ways of rendering the wax from the comb. I will describe two: the hot water method and the solar extractor. For the hot water method, place the wax in a cotton bag, take up any slack and tie it tightly closed. Immerse it in water in an old pan and arrange something to hold the bag well below the water surface, for example wire mesh, or a stone. Bring the water to a gentle boil. Simmer it slowly or hold it close to boiling for an hour or so. Leave to cool. The wax cake is lifted off the surface when it is set. The yield is fairly low this way. Depending on what you want to use it for, e.g. candles, you may need to hot filter the wax to remove particles. An alternative purification method involves settling particles after re-melting the wax. When sufficient wax has been collected, melt it in a container with an equal volume of water immersed in a bain-marie and let the whole cool very slowly without disturbance, e.g. in an insulated container. The particulates concentrate at the wax-water surface and can be scraped off the resulting cake with a knife. But if the wax is to be used only for waxing top-bars, this is not necessary.

I render my pressed comb in a home-made solar extractor. As pressing still leaves some honey in the comb and this clogs up the extractor over time, I crumble the comb cake in lukewarm water to dissolve residual honey, strain the liquid through a strainer and allow the comb to dry for a few days on a cloth supported on a plastic basket. The dry comb is left in the solar extractor for a day or two or until the last drops of wax have come out. My solar extractor has a cotton sheet under the comb contents so the wax emerges ready-filtered and suitable for use in candles and other beeswax products. Solar extractors are available from apicultural suppliers, but being expensive, beekeepers often make them themselves out of old double-glazing panels.[18]

---

[17] http://warre.biobees.com/pressing.htm

[18] http://www.dheaf.plus.com/warrebeekeeping/solar_extractor.htm

11   Feeding

Your policy on feeding will very much depend on your fundamental attitude as a beekeeper. If you have invested heavily in your bees, you are hardly likely to want to let them starve for want of a bit of honey or sugar. You may even feel a duty of care once you have bees in a hive more or less under your control. But overdoing feeding can remove a natural selective pressure that would otherwise act on the bees, thereby reducing fitness. One could envisage, for instance, that colonies which might not otherwise reproduce through being too improvident or not sufficiently thrifty, or are even a little diseased, are helped over a hurdle by feeding. You may therefore prefer to be cruel to the individual to be kind to the population. But if through long term poor foraging conditions most of your colonies need help to get through the winter, you will very likely feed them. Such a situation is hardly a fault of the bees. It will be a question of finding a balance that suits you. When supplementing winter stores, the best time to do it is in early autumn.

Firstly we consider the different types of feeder and then discuss which to use, when and with what. You could do all your feeding with screw-top jars and/or plastic food containers with tight-fitting lids. There is no need to go to great expense. However, if you have several colonies you may prefer the convenience of the specially designed feeder.

11.1 Feeders

11.1.1 Top contact feeders

These are especially for when there is a need to avoid breaking up the cluster, or when bees cannot even leave the cluster, for example in very cold conditions. Warré suggests modifying the top-bar cloth to contain a central hole covered with a wire gauze that can be closed with a flap of the cloth. He placed on it an inverted jar of syrup (honey or sugar) over the opening of which was firmly tied a cotton cloth. The bees lick the syrup from the cloth and the air gradually enters to release the vacuum. The advantage is that the feed may be replenished without letting the bees out. A snag is that the gauze limits the access of the bees to high capacity feeders or candy.

Instead, when feeding, I substitute another cloth with a central 75 mm (3") hole. Alternatively, cut a flap corresponding to a gap between top-bars. Over the opening I place a honey/jam jar or a larger

Fig. 11.1 Contact feeders of different capacities; note lid perforations

food container with a couple of dozen approximately 1 mm (1/16") diameter holes drilled or punched with a nail near the centre. The working principle is the same. The jar is housed in an empty quilt frame and usually surrounded with insulation, which may simply be a cloth bag containing hive quilt contents. With this arrangement, the roof rim is below the top-bar cloth so no rain can penetrate. If a larger box is used to house the feeder, and the quilt stored above it, the top-bar cloth rim is now exposed risking dampness entering. I find this is rarely a problem if the feeder is removed when feeding is complete. Taping the seam is another option.

For supplementary autumn feeding, the container may be as big as a 4 litre (gallon) food bucket housed inside a Warré box with no top-bars. As syrup is spilled while the vacuum establishes after inverting a feeder, use a spare closable container to catch the spillage, to avoid the risk of triggering robbing. Furthermore, the bigger the container, the more critical it may be that the hive is completely level. One wants to avoid the situation of syrup seeping out and eventually being accessible to robbers.

11.1.2 Other top feeders

An inexpensive high-capacity top feeder, sometimes called a rapid feeder, can be obtained from beekeeping suppliers. It is a translucent plastic circular trough with a central hole

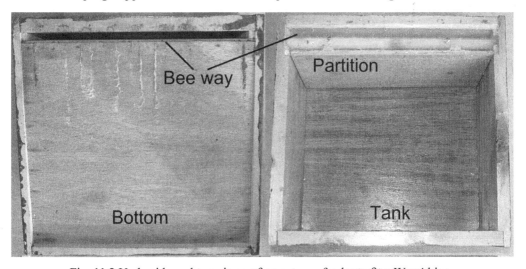

Fig. 11.2 Underside and top views of an autumn feeder to fit a Warré hive

and bee-way up which the bees crawl to drink without the risk of drowning in the bulk of the liquid. Whether it needs topping-up can be seen at a glance after lifting the lid, which is removable without disturbing the bees. This feeder fits a Warré box or quilt frame and may rest on the top-bar cloth with a hole cut in it.

For autumnal supplementary feeding Warré advises using a wooden trough top feeder the size of a Warré box (page 60, *Beekeeping for All*). It is based on the Ashforth feeder principle and holds about 10 litres (2½ US gallons). A strong colony can empty one of these in a day or two. The top-bar cloth is folded back just enough to expose the bee-way in the feeder. Refilling can be done without disturbing the bees. The advantage of Warré's design is that the floor slopes towards the bee-way so no bees can enter the tank until they have drunk all the liquid. At the time of writing, a Warré-size Ashforth feeder is available from Ickowicz, an apicultural supplier in France. However it can easily be made by any moderately skilled woodworker. The critical thing is to ensure the tank is well sealed. Some use beeswax for this, others varnish or paint, which have the potential disadvantage of releasing extractables into the syrup.

There is nothing against feeding in the old traditional beekeeping way at the top. A dish of syrup is placed in an eke of sufficient depth on the top-bar cloth, which must be folded back slightly to give the bees access. The dish is filled with straw or the surface almost entirely covered with wine corks to stop the bees drowning. Trickle a little syrup down into the nest to tell the bees it is mealtime. This is not a contact feeder, as the bees must leave the cluster to feed which they will not do in cold weather.

Fig 11.3 Bottom feeders; note viewing inside with mirror. Photo: John Haverson

11.1.3 Bottom feeders

Warré suggests his ingeniously designed bottom feeder for spring/summer use (pages 61-2, *Beekeeping for All*). It has a drawer that pulls out at the back of the hive to allow topping up while causing minimal disturbance to the bees. It is apparently not on the market, but is fairly easy to make. I have never made one, as I find the other options convenient, especially as I have a hive lift. Any sump or floor designed with rear access may be used in the same way (Fig. 2.25).

A bottom feeder may be any kind of container up to about 10 litres (2½ US gallons) either filled with straw or the surface covered almost entirely with wine corks. In my view the disadvantage of bottom feeding is the risk of debris from the brood nest falling into the feeder while in use. This potentially undermines the colony's hygiene precautions. Of course, the feeder could be designed in such a way that its liquid surface is protected by a cover.

11.1.4 Other feeders

An entrance feeder, available from apicultural suppliers, has the advantage of ease of installation on the outside, but the disadvantage that it places the feed outlet quite close to the entrance where it might attract robbers, especially if honey syrup is used. Also, with the Warré hive entrance, which has a step upwards just inside, insertion may be impossible without some modification to the entrance.

Plastic bag (e.g. ziplock) top feeders are popular with some, but as they are single use, on sustainability if no other grounds I see little to recommend them over closable rigid plastic or glass containers. Also they do not fall into the category of contact feeders because bees have to crawl onto the top of the bag to reach the feeding slits.

11.2 What to feed

By this stage of the book, the reader will have become familiar with how to encourage the bees to co-operate either with the help of a feather/brush or by using smoke. Both may be necessary when feeding colonies.

11.2.1 Syrup

The natural beekeeper no doubt prefers to feed honey. If it is given as syrup, in order to avoid the risk of fermentation, it should be prepared in small amounts that can be taken in a day. Two parts of honey are mixed with one part of water, both measured by weight. However, new beekeepers have none of their own honey, and bought honey, unless from an apiary very well monitored for foulbrood and other diseases, may contain pathogens. So the next best option is a syrup of *pure* organic sugar, from cane or beet, made from two parts of sugar to one of water, both by weight. Impure (brown) sugar places a serious burden on the bee digestive system, especially in winter when defecation flights are not possible. The main disadvantage of sugar is that it does not contain the micronutrients and microflora of honey, the proper food of the bee. Furthermore, excessive sugar feeding risks it being deposited in comb that is later harvested as 'honey'. Food dye tests have shown that, like honey, sugar syrup is moved around in the hive and can end up mixed in the honey.

Liquid feed can be used for newly hived swarms when the weather is poor, or for supplementary feeding in the autumn to prepare hives for winter. Natural beekeepers refrain from 'stimulatory' feeding with syrup intended artificially to induce early and rapid brood development in the spring, but if stores are dangerously low, feeding may nevertheless become essential. In which case use fondant or candy (§ 11.2.3).

Demeter/biodynamic beekeepers feed sugar syrup made with chamomile tea to which a tenth part by weight of honey and a pinch of salt are added. I have not tried this, but see no fundamental objection to it. Bees drink various natural 'teas' in the environment including guttation water, manure heap effluent, salty effluorescences on the soil surface and in other places, so why not herbal teas too? Bees readily take the herbal syrup.

11.2.2 Honeycomb

Honey in comb may be fed back to the colony by uncapping or scratching the cell cappings and propping the comb in a feeder that can be entered by the bees. Avoid allowing a situation to arise where the bees drown in seeping honey. If necessary use straw or twigs to create a scaffolding. Your extracted comb fragments may be cleaned dry by the bees in this way.

11.2.3 Candy

If for any reason winter stores have been underestimated, and the bees appear to be near starvation in the early part of the year, the remedy is to feed candy or fondant, its very close relative. Fondant, purchased from beekeeping suppliers or local bakers, is fed on a suitable support over a hole in the top-bar cloth. Here, I explain how to make candy. Its advantage is that it is a dense source of calories that the bees need to keep warm, and it can sit right on top of the cluster, requiring no bees to leave the cluster to feed.

Easy candy is made as follows: to a 1 kg (2¼lb) paper bag of sugar, add 120 ml (4oz.) cold water. Leave to stand in a dish overnight. Next day, close the bag and put a rubber band round it. Substitute for the top-bar cloth one with a 75 mm (3") hole in it, and, having cut a slit in one broad side of the bag, place the bag slit downwards over the hole. The bees will quarry the damp sugar, eventually leaving an empty bag. Surround the bag with insulation, e.g. hive quilt filling in a cloth bag (for ease of removal/replacement).

More complicated candy is made by completely dissolving sugar in a minimum of water, boiling until 117°C (243°F) is reached ('soft ball' on a cooking thermometer) and allowing to cool while stirring. When crystals just start to form, the mix is quickly poured into moulds to set. The candy is placed over a hole in the top-bar cloth for the bees to quarry it. Round that is a quilt frame and the rest of the quilt may be filled with quilt insulation, preferably contained in a cloth bag.

For a large slab of candy made from 4 kg (9lb) sugar, pour the crystallising syrup straight into an approximately 50 mm (2") eke/shim in which is seated a sheet of plastic resting on a baking sheet. To install it without letting out bees or nest heat, fold the top-bar cloth back about 50 mm (2") or less while sliding over the gap that opens a thin sheet of wood, plastic or metal. Put the slab of candy on top of the hive and slide out the baking sheet. Then slide out the sheet with which you have covered the aperture beside the top-bar cloth, thereby giving bees access to the candy. Replace the quilt and roof. As the top-bar cloth is now exposed round its edge and there may be draughts or rain seepage through the thin slit created, protect the joint with adhesive tape.

11.2.4 Dry sugar

I have not found it necessary to feed dry sugar. It is used as a late winter or early spring feed and would have to be placed in a feeder or in some sort of container in an eke above the bees, giving them access to climb up into the container to reach the surface of the sugar. Unlike candy, it has the advantage of needing little preparation apart from installing a feeder, but the disadvantage of not putting the feed in direct contact with the cluster. Lightly spraying the sugar with water helps to get the bees interested.

11.2.5 Pollen

I have never had to feed pollen to my bees, as it is available all the year round in my locality, e.g. from gorse (*Ulex europaeus*). However, even in parts of the UK, pollen dearth is occasionally a problem. Pollen is available from apicultural suppliers or local beekeepers. Beware of substitutes. Make thin pollen 'pancakes' by mixing it to a very thick paste with honey. Put the cakes between sheets of greaseproof or waxed paper over a hole in the top-bar cloth after first making some cuts in the lower paper to give bees access. There is no need to put an eke/shim round the feed in this case, but, especially if the quilt contents are quite heavy, put the lid of some container or a saucer over the pollen to prevent the quilt from pressing down on the feeding bees. The quilt cloth will be displaced a little upwards.

11.3 When to feed

I only feed in spring and summer when I have had to hive a colony in or just before a period of bad weather, and there are not already stores in the hive. In some areas though, bees can starve even in the summer, so it is as well to check stores or heft (§ 9.3) if you see that foraging is inhibited for a long period. My main supplementary feeding time is the second half of September to early October. It must occur before it has become too cold for the bees to take down the feed, store it, ripen it, and cap the cells. This could be in October, but sometimes ivy (*Hedera helix*) flows in warm spells continue into November. Over six bad summers since I started keeping bees in Warré hives, I have had to supplement stores in most hives in the autumn.

It is advised that when giving the autumn feed in an apiary, it should be done in the evening when there is less chance of triggering robbing. I have not found this to be necessary, although I have noticed a good deal of excitement around entrances shortly after putting feeds on hives. This probably results from bees returning to the nest from the feeder and doing the round dance that says there is feed in the immediate vicinity. Nestmates then frantically search for it. But they soon settle down and rush into the feeder. Of course, spillages should be avoided and covered up if they do happen.

Snowdrop (*Galanthus nivalis*) pollen loads. Photo: Alan Nelson

12   Wintering

The beginning of the wintering in my apiaries is marked by the end of supplementary feeding, removing surplus boxes, and putting mouse guards (§ 2.4) on. As mice have been known to enter hives in October in my locality, I put the guards on at the beginning of that month.

Fig. 12.1 Mouse guard secured with drawing pins

By this point of the year you will have hefted the hive, or counted combs of honey and left sufficient or all available stores for your bees. Warré advises wintering the People's Hive with two boxes full of comb, the top one almost full of honey and the bottom one undergoing its autumnal shrinkage of the brood nest, any additional boxes being removed. In my locality, the brood nest is free or almost free of brood by December, but depending on the race of bee and the climate, you could have brood even in winter. You may even see the odd bee returning with pollen in December or January. A very important feature of the bottom box is that it provides empty cells for the bees to cluster in as they eat their way upwards into the honey stores.

I myself and several others have successfully wintered small colonies in one box (Warré nucs). This situation could arise, for example, when a late swarm has been hived. The first one I had of this kind wintered on only 4 kg (9lb) stores and built up to a prosperous colony the following spring. The situation is little different from wintering a nuc for a frame hive, except that in mild, temperate climates like mine, the thermal properties of the Warré hive reduces the need for any extra insulation or sheltering. However, placing an empty box below the occupied box could help shelter the cluster from draughts from the entrance.

In much colder climates, two boxes may be insufficient as it may be necessary to have honey in the whole top box and in all or part of the next box down, in which case one is forced to winter the bees in three boxes.

In reducing to two or three boxes, what do you do with surplus boxes of comb, especially if you have a storage problem? Although I am in a climatically fairly similar region to Warré's, I ignore his advice and leave all third boxes in place that are all or partly filled with comb. I reason that the bees do not heat the space below the cluster, and that the extra box puts them further away from draughts from the entrance. If the hive is being disturbed anyway, this box, if free of brood, may even be set warm way to further reduce draughts. Remember to re-orientate it correctly in early spring!

The question of extra insulation comes up from time to time. I understand that wrapping hives or keeping them in a bee house is essential to their survival in the colder parts of Canada. Tar paper, a heavy-duty paper used in construction, is a suitable wrapping material. It is important not to overdo insulation. It not only cuts off the heat of the winter sun, but also risks *increasing* consumption of stores. Minimum use of stores occurs at about 5°C (41°F). Either side of this temperature honey consumption rises. Too much insulation, and stores are used faster not slower, because the bees are more active. Indeed they might even start raising brood, which quickly consumes the honey.

As mentioned in § 2.2.2, in extremely cold climates with snow on the ground for months it is prudent when making boxes to provide a small top entrance or closable flight hole in the box wall. The entrance may be opened only when there is a real need for it. With no top entrance, it is of course essential to keep the floor entrance clear of both snow and dead bees. A Warré entrance is more prone to such blockages as there is an alighting board where snow settles, and a small vestibule formed by the floor notch into which bees may drop and accumulate, especially when a mouse guard is restricting the entrance. Resting a slate against the front of the bottom box prevents snow settling on the alighting board.

The hive should not be disturbed in winter. Serious disturbance can disrupt the cluster and it may take a long time to re-establish itself, thus wasting stores and risking killing bees if they are so far out in the cold that they become too torpid to move back into the cluster. But on warm days in late winter, when the ambient temperature rises above about 8°C (46°F), if you have any reason to be concerned whether the stores are adequate to see the colony through to the first nectar flows (e.g. willow, *Salix*), you may heft the top box or even take a quick look under the back of the top-bar cloth. For emergency feeding at this time, I suggest candy/fondant (§ 11.2.3).

At other times you will content yourself by listening to the cluster by putting an ear to the hive or inserting the end of a stethoscope tube in the entrance. If you hear an unmistakable rustling-scratching-murmuring interspersed with the occasional slight buzzing, all is well. At each visit, be sure the mouse guard is still in place.

Calamities to watch out for, especially in winter, include wind toppling hives; trees falling on the apiary (it happened to me!); domestic livestock entering and scratching themselves against hives, and woodpecker, rat, skunk, badger, bear, etc. damage.

A particularly reassuring and pleasing sight in late winter is the first pollen coming in (*Galanthus, Crocus, Helleborus*, and, soon after, *Corylus*).

13   The spring visit and beyond

Of course, what Warré calls 'the spring visit', the time when you prepare the hive for the spring build-up, may be any one or more of the visits you make to your apiary in springtime. In my locality, the mouse guards come off in late March, when the bees are already very active and could certainly deal with a mouse if it dared to enter. Any dead colonies I detect by their silence as the winter progresses. Colonies that are alive but not bringing in pollen by the end of March may have a failed queen. If this continues for a week or so when other colonies are foraging, check for the presence of brood and eggs (§ 7.2). Bear in mind though that a colony in early spring with less than a palm-sized cluster of bees, yet with a laying queen, may expand to a prosperous colony and sometimes even give a honey surplus in the same season. I have seen it happen.

Any colonies with failing or non-laying queens you may wish to unite with stronger colonies (§ 7.8), thus freeing up equipment for more productive use. Occasionally a colony with a failed or no queen may become excessively defensive. Move it to the other side of the apiary, as far as possible from its stand. Old bees, those most likely to sting, return to the stand, so you can work without their interference. In such cases, because of the low bee population, it is relatively easy to find and remove the queen if present. If the colony appears to be queenless, it is probable that this has been so for some time, some workers having developed ovaries and started laying. This is evidenced by several eggs per cell, often on the walls. Drone brood from the sterile laying workers may already be evident. If so, instead of uniting by combining hives, which could result in the queen of the receiving colony being killed by the laying worker colony, take the hive 50-100 metres/yards away and brush all the bees from the combs onto the ground. Ordinary workers fly back to the apiary having little choice but to 'ask nicely' for admission to viable colonies. Laying workers having never flown, will not return. If a spot at a suitable distance is not available, given that the defective colony at this time of year is small, one may also do the operation in the apiary.

Close up the vacated hive and, until reuse, store it at a site different from the original. The comb may be rendered for wax, or, if free of disease, used for bait hives, or for repopulating a Warré hive. In the latter cases, first cut out any drone brood comb that may be present and destroy it. You may deal similarly with any comb from winter 'dead outs', only first free it as far as possible from any residual cluster of dead bees. It is worth keeping in mind that common beekeeping practices that involve transfer of comb from one colony to another are very prone to the risk of transferring pathogens at the same time. Therefore, it is as well to proceed with caution. If you are in any doubt, choose the more hygienic route of rendering the comb for wax. Instead, hives may be baited with pheromone and colonies may be established in either a new box or one that has been thoroughly scorched inside (§ 15.1). In my boxes with windows I have not had the glass crack when scorching. However it is advisable either not to play the flame directly on the glass or remove the glass beforehand.

Incidentally, I do not attempt to requeen my hopelessly queenless hives with a comb of brood. Firstly this is disruptive and weakening for the donor colony, and secondly, repopulating a hive with a swarm is so much easier. A swarm given a box or more of brood comb and some stores has a running start, so in my locality this can be done even up to the beginning of August.

I do the spring nadiring (§ 8) and floor clean in a warm sunny spell in the first half of April. The midday average temperature is then approaching 12°C (54°F). Although Warré says that it is permissible to add all the season's boxes at once, I add only one to each hive. I never seem to have enough boxes, so like to keep my options open with any boxes that are not immediately required. However, to vigorous looking colonies, you may wish to add two boxes immediately in order to avoid another intervention too soon. In any case,

Fig. 13.1 Disguised bait hive. Photo: Joe Waggle

maintain some space below the bottom of the comb so that lack of space, at least here, does not trigger the swarm impulse.

As the main flows will have not yet begun, while you are anyway disturbing the colony, you might as well check the stores at the same time, even though you may have done so earlier in the year. Either look under the top-bar cloth, or heft the top box. During the spring build up, honey/nectar consumption rises steeply. Some say 2.5 kg (6lb) honey should be on a hive at all times. On a Warré, this could be reduced to 1-2 kg (2-4lb).

Check the contents of the quilt and the top-bar cloth for dampness and renew as necessary. Warré suggests automatically renewing these materials. I rarely do so. My quilt contents are wood shavings and they seem to stay perfectly fresh smelling for years on end. Furthermore, the top-bar cloths are welded down with propolis, the bees' natural antiseptic sealant. It seems to me to work against the intentions of the bee to peel off the cloth and add a new one. Of course, if dampness has penetrated, for example when a feeder or feeding eke/shim has been placed on the top box, it is advisable to replace the quilt top-bar cloth which no doubt will have become mouldy.

Now is the time to ensure that any bait hives you require are placed and ready (§ 5.3). Monitoring for the rest of the season proceeds much the same as described in § 7, except that in your second season and beyond you will quite likely have the added enjoyment of dealing with swarming.

13.1 Forced comb change

If you feel that the top box is overdue for a comb change, for example if there has been too little honey on the hive to harvest the top box in previous seasons, there is a way that you can force the issue, assuming you consider the intrusion justified on hygiene grounds. This is not a procedure I do myself, preferring to allow natural development or colony failure to take its course. In the latter case, I can remove the comb and render the wax.

The comb change should be carried out in the spring when there is a good supply of nectar. It is at this time that the bees have the best conditions for coping with the intrusion. Before removing the top box, all the brood must be hatched out. To achieve this, the queen is driven down into the next box that contains brood and a queen excluder put under the box to be removed.

First free the box (§ 8.1) and smoke the bees down. Introducing light and smoke should drive the queen down. Remove the box and place a queen excluder on the hive. Replace the box and close up the hive. If you are in any doubt that smoking down the bees was inefficient, after three days have elapsed, examine a central comb or two in the top box for eggs (§7.2.2). If you see the queen on one of these combs, catch her in a queen clip, remove the excluder and put her below. If you see eggs, the queen is still in the top box. Remove the excluder and repeat the smoking down. Replace the excluder and leave the top box on it for at least 26 days. You may need to lift the top box now and then to allow drones to escape. This is best done on warm sunny days. The brood will hatch out and the empty cells may or may not be back-filled with honey.

You may leave the top box on to fill with honey like a super or remove it for harvesting and re-use. If you are concerned that the bees may not be able to do without the honey in the box removed, then it may be fed back to them from underneath the brood nest. In which case, first uncap the honey or scratch the cappings. With honey exposed at the bottom of the brood nest where there is greater risk of attracting robbers, consider whether you need to restrict the entrance. An additional precaution would be to insert this box in the evening when flight traffic has subsided.

14   Colony reproduction

Warré describes making increase by natural and artificial swarming (splits). Obviously, the most natural way is via swarms. In Warré's versions of artificial swarming, he advises using a mated queen. Thus splits could be made before occupied queen cells have developed naturally, and this would also avoid relying on the bees in the queenless half of a split to raise a new queen through building emergency queen cells. However, queen introduction is not without its problems and the increased cost of buying a queen might not be considered worthwhile for the month or so subtracted from the delay in egg-laying in the resulting split. Here I cover making increase by natural swarming and by splitting with normal (pre-swarm) queen cells present.

Making more colonies via splits provides an opportunity for breeding and selection. Natural selection does some of the work for you through natural failure of colonies. For example, before varroa spread to the western honey bee *Apis mellifera*, losses were said to be under a tenth of colonies, usually over winter, the most challenging season of the year for bees. But with varroa's arrival, annual losses greatly increased and with them natural selection increased in severity. Only the fittest survived the ravages of varroa. To some extent beekeepers undermine natural selection by feeding colonies that would not otherwise be able to provide for themselves, either because of pests or disease, or through some inherent weakness, for example, lack of thriftiness when forage is sparse. But despite artificial feeding, there are losses, so natural selection continues to some degree.

If you choose the less natural option of making increase by splitting, for example as part of swarm control, it makes sense to use your most vigorous colonies. That combines breeding with selection. I do not select individual traits, but one you may wish to avoid retaining, especially in an urban setting, is excessive defensiveness. However, this comes at a price. Defensiveness is correlated with fitness, so selecting against defensiveness to some extent thwarts *favourable* natural selection.

It is as well to keep in mind the possibility of inbreeding. Although queens are normally mated in drone congregation areas where much genetic mixing occurs, if you are in an isolated spot in a region poor in honey bees, you may need to consider bringing in 'new blood'. This need not be from very far away, and could be just the occasional swarm.

14.1 Natural swarming

The legendary Brother Adam (1898-1996) is not normally associated with natural beekeeping, but I would like nevertheless to share this quote from his book: 'There can be no doubt whatever that the swarming impulse provides the best nurtured and best developed queens, for when a colony prepares to swarm it has reached an optimum in its organic development, as well as opulence in every direction. Indeed swarming is the natural manifestation of a colony having reached the summit of affluence. In such circumstances ideal conditions prevail for raising the best of queens from a physical point of view'.[19]

---

[19] Brother Adam (1986) *Beekeeping at Buckfast Abbey*. Northern Bee Books, Mytholmroyd.

In *The Bee-friendly Beekeeper*, I discuss some of the scientific evidence in favour of letting colonies reproduce by natural swarming. In brief: the swarm leaves behind the bulk of the pathogens, is normally subject to natural selection in its speed and efficiency of finding shelter, and comes ready fuelled and eager to build a new nest. On the other hand, the beekeeper must be very aware of events at the apiary, the apiary location may be far too urban to allow swarm issue, and, in the case of a prime swarm, the queen is already a season or more old. I make increase in my colonies or replace dead outs with natural swarms. I deal with taking swarms in § 5.1. Always have your swarm kit and a spare hive at the ready!

I am fortunate in living in the countryside and having the time to visit my apiary daily in the swarm season (May to July), sometimes several times a day. To help minimise swarm loss, or merely to alert me to the fact that a swarm issue is imminent or already issued nearby, I use bait hives. Half a dozen scouts seen at a bait hive is the signal to pay closer attention to the apiary nearby. Perhaps a swarm is already hanging in a tree. A few dozen or sometimes hundreds visiting at a bait hive usually means it has been chosen by a swarm on the move. For hiving or transferring swarms see § 6.1.

It is sometimes said that it is irresponsible to allow swarms to issue naturally, because it risks them settling in buildings, thereby creating a nuisance and possibly an expensive job of removal. This argument has only become feasible because of the superior swarm suppression offered by the ease of inspection, and subsequent manipulation, due to frames. However, this suppression is a relatively recent development in the millennia of beekeeping. Formerly, swarms of bees were accepted and tolerated as natural phenomena. Yet despite the frames, swarms do get away. Indeed, three hot spots to which I am called to remove colonies from buildings are within a few hundred metres of large apiaries of experienced beekeepers running frame hives. Diligent natural beekeepers, rather than believing that their colonies do not swarm, as do some frame beekeepers I have met, will not only welcome swarms but also try to minimise their impact on others. Ways of doing this include strategically placed bait hives, alertness to events in the apiary, and asking people roundabout to tell you if they see a swarm. You then have an opportunity to take it before it settles in a cavity.

Fig. 14.1 Scouts at a rooftop bait hive

Swarms are not only valuable for making increase or replacing winter dead outs but also, even when only casts (secondary/after swarms), as an insurance policy against late season queen failure. A cast could be housed in a single Warré box, making a Warré nuc. If you are short of equipment, an improvised and temporary quilt, roof and floor could be used. A late cast could be fed honey, if forage is short, and kept on one side in reserve. Towards the end of the season, if it looks as if you are not going to need it, it might be of use to another beekeeper near you. If you reach a time when you think you will give away any further swarms to emerge, alert other beekeepers so that they may have hives at the ready.

Fig. 14.2 Strange swarm: after issue it eventually clustered on the ground
round a small stick. It was hived by placing over the swarm a hive
supported on two sticks and with the floor removed.

14.2 Simple split

If for any reason you feel obliged to prevent a Warré colony from swarming, thus denying your colonies the benefit of natural reproduction, you have the option of making a split and, if longer term increase is not desired, later reuniting the two halves. This is best done at the start of or a little before the swarm season in your area, when the colony is already prosperous enough to be swarm-ready. Indeed, it may already have occupied queen cells. An outward sign of the swarm mood is bearding at the entrance. When it comes to searching comb by comb for the queen, you may regard as a daunting challenge a People's Hive with top-bars and three or more boxes of brood in a state of readiness for swarming. Of course, if you have opted for a frame- or semi-frame version of the hive, the task would be easier. You could use any one of the various artificial swarming methods described in the frame hive literature. One is presented below (§ 14.3), only slightly adapted to the case of comb on top-bars. Colonies started from splits are not infallible, but then neither are those started from natural swarms.

But whether your hive is frame or top-bar, you may carry out a split without finding the queen. You will require an empty complete hive with prepared top-bars. The parent colony should be strong and in no less than three boxes. Place beside the parent colony a floor and on it one box with top-bars. Open the parent colony, remove the top-bar cloth and smoke between the top bars. The smoke and letting in the light is to drive the queen down into the lower boxes. Remove the top box (§ 8.1) and inspect underneath it. If occupied queen cells are visible, you have a colony that is already rearing new queens in the natural way. At the very least, you need to be certain that the top portion of the brood nest is visible in this box. If the box is entirely devoid of brood you must either abandon the split or, if it is a 4-box hive that is being split, try to make the split between boxes 2 and 3. Place the box on the prepared empty box in the same orientation with respect to the entrance as it was on the parent colony. Put the top-bar cloths and quilts on both hives. Move the parent hive with the queen to a distant point in the apiary or, later in the day, to another apiary. Move the new hive containing brood but no queen to the former position of the parent hive. This will now receive the field (forager) bees and strengthen the new colony, which should go on to raise a new queen. She should be laying in about four weeks. If after putting the new hive on the site of the parent colony, there is not sufficient space to move the parent colony elsewhere, erect some sort of barrier (branches, windbreak fabric etc.) between the two hives so as to force a distinct separation in flight paths of returning bees and minimise their drifting back into the parent colony.

14.3 Artificial swarm

As artificially swarming is sometimes essential as a means of disease control in some states/countries, for example if the colony succumbs to European foulbrood, it would be remiss to omit artificial swarming from a book such as this, even if the reader never intends to split colonies this way for the purposes of making increase. For disease control only, the artificial swarm is normally done by shaking bees off comb. The combs are then destroyed. But it is not possible to make a 'shook swarm' with a Warré hive colony because shaking would detach the combs from the top-bars. So the remaining options to achieve the same ends are smoking and brushing the bees. Admittedly, there is the possibility of driving the bees (§ 6.3.2.1), but this seems to have gone out of fashion.

The aim is to transfer all or most of a colony in three boxes, including the queen, into a new hive and allow the nurse bees to repopulate the parent hive through a queen excluder and thereby cover the brood. The transfer may be done without finding the queen and by smoking or by brushing, or even sometimes by a combination of the two. For smoking, you will require an empty complete hive with prepared top-bars and a queen excluder. For brushing you will need the same plus a box without top-bars, a floor (however improvised) and a cover which may be simply a top-bar cloth.

Making colony increase this way should not be undertaken lightly as there is a period during it when the brood is free or almost free of a cover of bees and therefore more liable to chilling. So choose a really warm sunny day. It will certainly help you if the foragers are out in the field.

14.3.1 Smoking method

Nadir a fourth box with top-bars under the occupied boxes, if necessary, at the same time clean or replace the floor. By now you are familiar with removing boxes from an occupied hive (§ 8.1). Remove the roof, quilt and top-bar cloth of the parent colony. Vigorously smoke the bees down into the second box. Remove the top box and set it aside with a temporary cover on a box stand. Proceed the same with the second and third boxes in turn, each time setting them aside under a cover. If the bees have responded well to the smoke, by now the artificial swarm and queen should be in the new (fourth) box. Nadir a fifth box with top-bars. Smoke or brush any bees from the top-bars/rims of the fourth box and put on it a queen excluder. Replace boxes 3, 2 and 1 in that order and in the same orientation with respect to the entrance as before. Go to § 14.3.3.

14.3.2 Brushing method

If, as sometimes happens, most of the bees and the queen are not driven down by smoking, you may have to resort to the more laborious and drastic method of removing the combs one by one (§ 7.2.2) and brushing the bees off. In which case, an assistant would be almost indispensable: one person removes combs and the other brushes bees off. Completely separating bees and comb is anyway essential if the artificial swarm is being done to control disease and not to make increase.

As making increase this way leaves the brood uncovered for longer, it is all the more important to do it in warm, windless weather. Furthermore, the increased period of the exposure compared with the smoking method suggests that confining the manipulation to a one- or two-box colony is advisable.

Have to hand an empty box with no top-bars on a spare floor and a temporary cover such as a top-bar cloth. Set aside the parent hive on a stand and place two boxes with top-bars in its stead, if necessary first cleaning or replacing the floor. Set on them a funnel or a third box to act as a funnel. Remove combs from the top-box of the parent colony and brush the bees into the funnel/box. Avoid damaging any queen cells present. Put the bee-free combs in the spare box and keep them covered to minimise chilling the brood. As you proceed, retain the original order and orientation of the combs with respect to the entrance. Pre-marking the top-bars of the box from which combs are removed with a diagonal line with a felt pen or a scratch with the hive tool helps with this. The box vacated of comb receives the combs taken from the box that was below it. When all the combs are brushed, remove the box/funnel, smoke or brush any bees from the top-bars/rims of the fourth box and put on it a queen excluder. Set on the excluder the boxes of comb from the parent in their original order, and in the same orientation with respect to the entrance as before.

14.3.3 Conclusion of both methods

Close up the hive and leave it for at least two hours for the nurse bees to move up onto the brood. Move the parent hive, now with its combs covered again with bees, onto a floor and stand at a distant point in the apiary or, later in the day, to another apiary. The new hive will now receive the returning field (forager) bees which will strengthen the artificial swarm. If there is insufficient space to move the parent colony elsewhere, erect some sort of barrier between the two hives so as to force a distinct separation in flight paths of returning bees and minimise their drifting back into the parent colony. As this process has been highly disruptive to the bees, there is all the more risk of the swarm absconding. You may wish to avoid this risk by placing a queen includer under the colony when the manipulation is complete (§ 6.1.3).

The resulting artificial swarm, unlike a natural swarm, will not necessarily have prepared itself by taking in a copious supply of honey. It would therefore be advisable to feed it fairly generously, especially if foraging conditions are not at their best.

15   Pests and diseases

My impression is that, relative to bee biology and behaviour, a greater part of the modern apiological literature is devoted to bee pests and diseases. If it is true, it is no doubt due to the economic impact of these maladies on apiculture influencing the flow of research funding. Keeping bees in a more natural way, if it *diminishes* the risk of disease, certainly does not *avoid* it altogether. Indeed, a look into the more accessible literature from the heyday of skep beekeeping, when the conditions in which bees were kept were arguably somewhat more natural than is largely the case nowadays, shows that the common bee diseases were known then, even if less was understood about them compared with today. To those maladies, we can now add the varroa mite which has reached most parts except Australia at the time of writing. Varroa, with its tendency also to spread viruses, is now generally ranked as the organism having the worst economic impact on bees. So we need to consider what we can do to minimise the effect of pests and diseases on our bees over and above the measures that are already built-in to the beekeeping methods that are described above. For a disease to arise, two conditions are necessary: both a susceptibility and a pathogen must be present. But neither alone is sufficient. The way we keep our bees can help avoid bringing those two conditions together.

15.1 Hygiene

It has long been recognised that beekeeping practices such as having too many hives in one place or moving from one colony to another equipment, combs (with or without brood), honey and bees, and so on, increase the risks of disease in that they undermine the hygienic processes built-in to the bee population. To that, we can add chilling brood and malnutrition. We can try to minimise such contributory factors. For example, we could make sure that all our colony reproduction is only into a new hive. But this is somewhat extreme hygiene and a compromise may be necessary, even if only to match the financial resources available.

Keep in mind what manipulations could transfer pathogens between colonies. Your hive tool and gloves, if waterproof, may be cleaned in a solution of washing soda (sodium carbonate). With a bit of rubbing, it removes propolis. A more effective way of sterilisation of the tool is by putting it in a hot flame: open the smoker, work the bellows until a flame is produced and hold the tool in it for a few minutes. If you have several apiaries, you could keep a separate hive tool in each, for example in the top of a hive quilt.

Floors and boxes may be routinely scorched with a propane torch before re-use. Top-bars may be re-used in the way Warré described, i.e. leaving a starter strip of 5 mm (¼") of comb on them only if the hive they came from was free of clinical signs of disease. If you are in any doubt, the bars may be sterilised in an oven at 140°C (284°F) for 90 minutes and, when cool, re-waxed. But if you have had *clinical* American foulbrood in a hive, check what your local regulations are for sterilising the equipment. For example, in the UK, the hive boxes would be scorched, and frames (and comb) burnt. The same would therefore apply to top-bars from an infected Warré hive.

In the extreme event of a colony having to be killed, for example on the order of a foulbrood inspector in your area, or because it is so diseased that saving it is impossible and there is a risk that surviving bees will spread disease elsewhere, there are three options. In the case of an order to kill a colony, often done under supervision, the advice of the inspector should be followed. Otherwise there are normally three fairly convenient possibilities: sulphur, petrol or a strong solution of detergent such as partially diluted dishwashing liquid.

To kill the bees by a method that does not leave potentially toxic residues in the hive, close up the entrance, open the top, remove the top-bar cloth and put on an extra box with no top-bars. Place burning sulphur in a metal container in the box and close up the hive. Sulphur is available from beekeeping suppliers. Be aware that powders comprising nearly pure sulphur available from garden centres may contain other substances used as wetting agents or dispersants. For a less 'clean' method, a rag soaked in petrol (gasoline) may be placed on a mesh over the top-bars and the hive closed up. The messiest method, though an effective one, is to douse/spray all the bees with a strong solution of detergent. Happily, I have not yet had to kill a hived colony, but have used detergent on a few colonies in buildings where removal alive was not feasible. The advantage is that, as the method involves no insecticides, it leaves no intensely toxic residues that might be spread by robbers to other colonies in the area.

15.2 Insects

A healthy Warré colony should be capable of defending itself against wax moths (*Galleria mellonella; Achroia grisella*), common wasps/yellowjackets (*Vespula vulgaris/germanica*) and even small hive beetles (*Aethina tumida*) . If a wasps' nest is near the apiary or if there has been a surge in the population in a particular year, in late summer it may be necessary to restrict the hive entrance while keeping it compatible with foraging traffic. You may wish to destroy any nearby nests that you can find. However, bear in mind that the wasp too has its beneficial aspects as it consumes aphids. In wasp plague years, near my hives I place wasp traps which, in their simplest form, comprise a glass jar with a small hole in the metal lid and a liquid containing jam in the bottom.

I find that wax moth is only ever a problem in unoccupied hives. If you must keep comb for reuse the following year, storing it for two days in a freezer should eliminate any moth eggs. The comb, or whole boxes of it, is then stored in a plastic bag.

The small hive beetle is a more serious problem. Early indications from Australia show that Warré colonies cope well with the beetle. With the hive's simpler internal architecture, there are fewer places for the beetle to hide. If you find that the beetle is a problem in your apiary, consider fitting an entrance trap of the kind available from apicultural suppliers. There are also floor traps available. To fit them on a Warré hive may require the alighting board to be removed.

The European hornet (*Vespa crabro*), is not normally regarded as a problem for bee-keepers. I have only ever seen one at my hives. However the Asian hornet (*Vespa velutina nigrithorax*) which, at the time of writing is advancing through France, and is expected to cross into the UK before long, is a much more serious pest to beehives.

15.3 Slugs

For a reason that I have not been able to fathom, my Warré hives are sometimes attractive to large slugs, occasionally as many as four. This never happened with my frame hives, which all happen to be on mesh floors. Perhaps the metal mesh is a deterrent. The slugs are not a threat to the bees, but they do sometimes leave faeces in the hive wall or floor. The bees rarely pay attention to them. I have seen only two slugs killed in a hive, and embalmed in propolis.

15.4 Varroa

By far the commonest bee disease I see in my colonies involves deformed wing virus. The most obvious outer symptom is feeble-looking bees with shrivelled wings crawling out of the entrance or on the ground. Occasionally they may be seen being dragged out by healthy looking worker bees. The virus is one of many that were present in bees before varroa arrived, but which are now spread by the mite as it feeds on the haemolymph of pupae and adult bees. But bees are not entirely defenceless against Varroa. Mite reproduction is inhibited by the broodless period of natural swarming, a warm nest environment, hygienic behaviours, and certain chemicals emitted by pupae. The superior thermal properties of a Warré hive should help create a warmer nest environment and inhibit varroa, but this has not yet been verified.

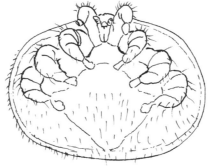

Fig. 15.1 Varroa mite underside x 50

The only treatment is to reduce varroa numbers. Many if not most natural beekeepers prefer not to treat for varroa. They are prepared to accept the higher colony loss rate in exchange for not having to put chemicals in their hives. I take this line with my own colonies, but I sustain Warré colony losses averaged over five years of about 30%. If a beginner has invested heavily in a hive, equipment and bees, they may not wish to see their investment destroyed by varroa/viruses. In which case they may consider using the treatments available and legally acceptable in their own locality. Your local association will be able to advise about this. Before 2009, when I was still treating for varroa in frame hives, I used oxalic acid at the beginning of January and thymol in September, after the honey harvest. Convenient preparations and delivery systems for these two active substances, and others such as formic acid, are available from apicultural suppliers. However, none of these substances is without its negative impact on the bees and the beneficial microbiota in the hive. Furthermore, any intervention against varroa may slow the process of natural selection which tends towards co-adaptation of bee and mite. It is a case of balancing cost with benefit.

Some natural beekeepers treat with powdered sugar, sold as icing sugar in the UK. It is said to be the most benign treatment. However, as there is good evidence to show that it is not effective, even when done weekly or fortnightly, and is very disruptive through having to open colonies very frequently, I do not discuss it here.

If you plan to treat for varroa it makes sense to treat only if it seems really necessary. A way to assess the mite burden in the colony is temporarily to substitute a mesh floor for Warré's floor. A drawer placed under the mesh and insertable from the rear holds a sheet of greaseproof paper or card coated with something that the mites will stick to, such as vegetable oil. The unit is left in place for three days and the natural mite drop counted. Depending on the time of year, the number dropped per day is multiplied by a factor previously found by experiment, in order to give a rough estimate of the number of mites in the colony: spring/autumn x 100; summer x 30; winter x 400. Treatment is advisable only if the mite count approaches the economic threshold of about one eighth of the bee population. The average number of adult bees in a healthy, vigorous colony may range from less than 10,000 to nearly 40,000 in summer.

Some organic beekeepers consider that 'retrogressing' their bees onto small-cell comb by the use of wax/plastic foundation or plastic comb with a cell size of 4.9 mm instead of 5.3/5.4 mm helps their bees cope with varroa. I have reviewed the experimental evidence for and against this hypothesis and, at the time of writing, have not found it sufficiently convincing even to attempt this treatment.[20] My bees build their cells to the size they choose. The term 'retrogressing' assumes that, before commercial foundation was available, bees were once on cells of 4.9 mm diameter. I have reviewed the historical evidence from apiologists right back to Swammerdam, and can find nothing to support the assumption.[21]

15.5 Diseases involving micro-organisms

The Warré hive, with its superior thermal properties, near-natural comb, and management involving minimum intervention, together with a built-in comb renewal process, is able to help minimise the kinds of physiological stress that precipitate some of the diseases associated with micro-organisms described below. If attention is also given to good hive siting and the colony is never short of a diverse natural nutrition, the natural beekeeper will have done all he or she can to minimise such diseases. Prevention is better than cure!

15.6 Nosema

If the entrance and alighting board is smeared with a lot of bee faeces, i.e. yellow or orange or brown deposits, suspect dysentery due to nosemosis. Bees crawling on the ground are another symptom. As this disease, involving the microsporidians *Nosema apis/cerana*, is largely triggered by stress, it is unlikely to occur with the bees in your care. But if it does occur, and you very likely do not want to use the recommended

[20] Heaf, D. J. (2011) Do small cells help bees cope with varroa? – A review. *The Beekeepers Quarterly* (June 2011) No. 104, pp. 39-451.

[21] Heaf, D. J. (2013) Natural cell size. http://www.dheaf.plus.com/warrebeekeeping/natural_cell_size_heaf.pdf

treatment with an antibiotic, see if reducing stress by feeding or even re-siting the hive at a sunnier, drier, better sheltered apiary will help. If the new site is away from other hives the added advantage is it avoids the risk of infected bees drifting to other colonies and carrying *Nosema* spores with them. But if the colony does not recover, you have little choice but to kill it (§ 15.1), scorch the hive insides, and start again.

15.7 Brood diseases

If the foulbrood diseases have become so advanced that you can detect them by entrance smell – old woodworking glue smell for American and sour smell for European – the disease is likely to be very far advanced and no remedy is possible, other than destroying the colonies. To inspect for brood disease, proceed by removing combs (§ 7.2.2). The beginner may wish to enlist the help of someone who has seen foulbrood. That may be the government's district inspector for bee disease or a member of the local beekeeping association. Beekeeping books, free publications from the UK National Bee Unit and from extension and governmental services (e.g. USDA) in other countries, together with many authoritative pages on the internet, offer a wealth of colour photos of brood disease.

15.7.1 European foulbrood

Larvae die and decompose at an early stage, leaving a spotty pattern in capped brood. Nurse bees try to remove them, thereby spreading the spores to other larvae. At the time of writing, it is a notifiable disease in the UK, although consultations are in progress which might change its status. Its treatment in the UK involves transferring the bees to a fresh hive, destroying combs, and scorching equipment. However, as it is considered a disease of stress (moving hives, chilling, malnutrition) it may come and go in a colony as nutritional and weather conditions change, and often without the beekeeper realising it is there.

15.7.2 American foulbrood

In this disease, notifiable in the UK and many other countries/states, the larvae die and decompose after the cells are capped. The outward sign is sunken cappings, sometimes perforated. On uncapping and dipping a matchstick into the gel at the bottom of a cell, a brownish gluey 'rope' forms between the cell and the stick. This gel contains millions of spores. It dries out and becomes difficult for the cleaner bees to remove. The spores are incredibly long-lived and heat-stable. In the wild, if the colony dies out, the decomposers often shred the comb, thus removing the source of infection. But if a colony enters before this, the disease will very likely continue in its brood. Faced with clinical foulbrood in a hive, whatever other options there may be, the natural beekeeper will no doubt at least destroy the comb, top-bars and top-bar cloth, followed by scorching the hive boxes and floor.

Apiologists consider American foulbrood to be a disease enhanced by beekeeping practices, including colony stress, and moving hive parts and comb between colonies.

15.7.3 Chalk and sac broods

These are largely stress diseases which the bees often remedy themselves. Occasionally,

I see chalk brood 'mummies' at my hive entrances. A fungus grows on larvae, usually on only a few here and there, and eventually nearly fills the cell. The bees can generally remove intact the resulting lump, which is oval, usually tapers towards one end, and is mostly whitish, though often with darker markings ranging to black. In severe cases, remediation by requeening may be necessary.

Sacbrood involves a virus and results in larval death just before or after cappings. Dead larvae look watery and granular in a thick sac with the head towards the cell entrance. If there are many dead larvae there is a sour smell. There are no treatments for either 'chalk' or 'sac' other than relieving physiological stressors. Requeening may help.

As one might expect, the list of honey bee parasites and pathogens is far longer than those I have given here which are only the more common and symptomatologically obvious ones that you are likely to encounter.

15.8 Pesticides

This is a vexed issue with beekeepers, and has been for many decades since the introduction of DDT. If you live in a largely grazing landscape, as I do, the risk of pesticide exposure is probably lower than in more arable landscapes. Even so, it has been suggested that even sheep dips may spread pesticides to meadow flowers through run-off from the animals.

The main pesticide threat at the present time appears to be neonicotinoids, which are widely used on arable crops and are quickly lethal to bees at very low doses. Chronic exposure to even lower doses can disrupt bee behaviour such as homing performance. Tens of thousands of colonies were lost in Germany in recent years due to drift from neonicotinoid seed treatment. The pesticide is taken up by the seed and distributed through the whole plant and may be transferred to bees through pollen and/or nectar.

Apart from lobbying political representatives for more bee-appropriate pesticides, or even for organic, pesticide-free agriculture, what is to be done? The first step would be to identify at an early stage in the growing season any crops in your locality that are likely to present a threat. Farmers have to comply with pesticide application instructions prescribed in licenses for the pesticide. That may involve them liaising with individual beekeepers and/or local beekeeper associations and avoiding spraying when the bees are out. It may be possible to close up hives during spraying, in which case it would be advisable to install a mesh floor beforehand and, if it is warm sunny weather, consider providing top ventilation. Erecting shading for the hive may also help. If the worst comes to the worst, you may need to find another location for your apiary.

Poisoning usually occurs in the spring or summer and is evidenced by many dead bees suddenly appearing round the hive entrance. If you suspect that your bees have been poisoned by pesticide, you may have a right to compensation, especially if the pesticide has been used irresponsibly. Collect, identify, and freeze samples of a few hundred bees from each of the hives affected and take photos, time-marked if you have a digital camera. Your local association will tell you which authority to contact. In the UK the National Bee Unit is one source of help.

Appendix 1
Web sites and forums on Warré and natural beekeeping

Warré Beekeeping http://warre.biobees.com

Yahoo e-group: http://uk.groups.yahoo.com/group/warrebeekeeping

David Heaf's website: www.bee-friendly.co.uk

Friends of the Bees: www.biobees.com

Natural Beekeeping Trust: www.naturalbeekeepingtrust.org

Gareth John's natural beekeeping blog: http://simplebees.wordpress.com

Nick Hampshire's site with Warré hive construction help for complete beginners: www.thebeespace.net

Jean-François Dardenne: www.ruche-warre.levillage.org

Guillaume Fontaine: www.apiculture-warre.fr

Jan-Michael Schütt & Olivier Duprez: www.ruchebio.com

Gilles Denis, commercial beekeeper, supplier of modified Warré hives, parts, bees and Warré beekeeping manual; runs courses on the Warré hive: www.ruche-warre.com

Marc Gatineau:
www.perso.orange.fr/marc.gatineau/ or http://www.apiculturegatineau.fr

Jérôme Alphonse: www.mielleriealphonse.com

French language forum: www.ruchewarre.net

French language Google e-group (Google membership required):
http://groups.google.com/group/la-libera-abelo?hl=fr

Claude Bralet's extensive site on ecological, bee-friendly beekeeping including the Warré hive: www.la-ruche-sauvage.com/ruches/rucecolo.php, or www.la-ruche-sauvage.com and navigate

Appendix 2
Warré hive suppliers (August 2013)

The quality and specification of hives sold as Warré varies a lot. Try to find out if the specification is what you expect. Take time in buying, and consider checking with several sellers. If you are in doubt, ask in Warré e-groups/forums if anyone would like to comment on a product that interests you.

USA
Chris Harvey (Michigan): www.thewarrestore.com (part of Teakwood Organics)
Dale Hill (Washington State): www.farmgardenandbeyond.com/bee-hives (also makes hive lifts)
Tel Jensen (Washington State): www.pikkufarm.org
Matt Reed (Oregon): www.beethinking.com (also stocks Warré comb knife)
Bill Wood (Oregon): www.beeologique.com

UK
Robaire Beckwith (Denbighshire): www.beesandthings.co.uk
Neil Cruickshank (Penrith): www.edenbeehives.co.uk
Mike Hillard (Stroud): www.tranquilityhouses.com/lifestyle/hives/hive_info.html
David Johnson (Penzance): www.natural-beekeeping.co.uk/buy-a-warre-beehive/
Matt Mercy (Dorset): www.zorbanet.com/warre/
Matt O'Callaghan (New Forest): www.warrebeehive.co.uk
E. H. Thorne Beehives Ltd. (Lincs., Hants., Berks., Fife) www.thorne.co.uk

European continent
Import-Handel-Export Alfred and Anita Thuminger, Vienna. (Delivers to the UK)
http://massivholz-tischler.de.server963-han.de-nserver.de/de-en/
(with or without a window in each box)

Ickowicz (France) www.ickowicz-apiculture.com/

Jérôme Alphonse (France) www.mielleriealphonse.com

Andreas Meisl (Germany): www.oekobeute.de (download PDF-Katalog, order by email)

Heinrich Holtermann (Germany)
www.holtermann-shop.de/index.php/cPath/1_234/category/warré-beute.html

Australia
Tim Malfroy (NSW): www.naturalbeekeeping.com.au

If any of the above links fail, please see http://warre.biobees.com/links.htm

Index

Glossary

Absconding Whole colony leaves the hive

Acaricide Mite killer

Anaphylaxis Severe immune reaction that can result in death.

Anthropocentric Man centred

Apicentric Bee centred

Bain-marie Hot water bath

Biodynamics Farming system indicated by Rudolf Steiner

Burlap Woven jute cloth

Cast Secondary swarm or after-swarm

Centrifuge Machine to spin honey out of framed comb

Cold-way Combs at right angles to hive side with entrance.

Crown board Hives internal cover board

Demeter Certification name of biodynamic produce

Drifting Bees returning to the wrong hive

Drone Male bee

Ecocentric Centred on ecological principles

Eke Hive body element, usually shallow, used to extend the hive to create extra space

Fondant Soft sugar candy used by bakers

Formic acid Acaricide naturally occurring in ants, bees etc., now synthetic

Foundation Wax sheets embossed with cell hexagons

Hefting Weighing, usually by feel

Hessian Woven jute cloth

Honey gate Tap used on ripener to allow bottling with minimal drips

Langstroth USA inventor of frame hives with bee space

Larva Grub stage of young bee from egg hatching to changing to pupa

Melliferous Honey bearing

Microbiota Totality of micro-organisms in a given place

Nadir Hive body box placed under the brood nest

Nasonov Name of gland on tip of honey bee abdomen; named after its discoverer

National Popular hive in the UK

Nestduftwärmebindung Original Thür term for retention of nest scent and heat

Nuc Abbreviation for nucleus colony/hive

Oxalic acid Acaricide found in e.g. rhubarb, now synthetic

Pathogen Organism associated with a disease

Pheromone Chemical signal from living organism

Propolis Wax and plant exudate mixture used as bee's natural antiseptic and sealant

Pupa Passive chrysalis stage of young bee after larva, in cocoon in capped cell

Quilt In this context: shallow frame filled with insulating material

Rebate (rabbets) Indentations in edges of hive body box sides to support top-bars

Ripener Settling tank with tap for allowing air and wax/propolis particles to clear from honey

Shim Spacer to separate two parts of something constructed. In a hive it is called an 'eke'

Super Hive body box placed over the brood nest

Supersedure Replacement of queen without swarming

Thixotropic Property of gels that flow when stirred but become semi-solid on standing

Thymol Acaricidal essential oil from thyme, now synthetic

Varroa Mite that lives on honey bees

Vaseline Petroleum jelly

Warm-way Combs aligned parallel to hive side with entrance

Acknowledgements

I thank Jeremy Burbidge of Northern Bee Books for suggesting that I write this book. Jan Jenkins and John Haverson kindly read the manuscript and made many helpful suggestions that have greatly improved the book. I thank my wife, Pat, for disentangling some of my sentences and preventing typos and other anomalies finding their way into print. Any errors or omissions that remain are my responsibility.

Although the book owes a lot to Abbé Émile Warré, who died in 1951, it would not have been possible so thoroughly to place his beekeeping method in a modern context without the reports from beekeepers around the world of their experiences with Warré's 'People's Hive'. In this respect, I would like to thank especially Andy Collins (UK), Dav Croteau (USA), Karman Csaba (Romania), Jean-François Dardenne (Belgium), Larry Garrett (USA), Guillaume Fontaine (France), Marc Gatineau (France), Jean-Claude Guillaume (France), Steve Ham (Spain), John Haverson (UK), Raimund Henneken (Germany), Bernhard Heuvel (Germany), Andrew Janiak (Australia), Johan Kok (UK), Timothy Malfroy (Australia), John Moerschbacher (Canada), Syouichi Morimoto (Japan), Alex Templeton (USA), Dietrich Vageler (Brazil), Bill Wood (USA), and many others who have contributed over the years to our knowledge of natural beekeeping with the Warré hive. Finally, I warmly thank Phil Chandler for donating space at biobees.com for furthering the Warré project.

About the author

David Heaf was born in 1947 in Liverpool, grew up in Sheffield and obtained his B.Sc. and Ph.D. degrees in biochemistry at the University College of North Wales, Bangor. After a career in research biochemistry he settled in north-west Wales where he works as a translator and, with his wife Pat, manages a large vegetable garden. He started beekeeping in 2003 with National hives. In 2006 he was introduced to the Warré hive and populated six of them in 2007. He winters about 15 Warrés and less than a handful of Nationals. In 2010, Northern Bee Books published his book on sustainable, apicentric beekeeping entitled *The Bee-friendly Beekeeper*. It was reprinted in 2012.

Lightning Source UK Ltd.
Milton Keynes UK
UKOW07f0229260815

257499UK00006B/182/P